NEW APPROACHES TO THE ECONOMICS OF PLANT HEALTH

Wageningen UR Frontis Series

VOLUME 20

Series editor:

R.J. Bogers
Frontis – Wageningen International Nucleus for Strategic Expertise
Wageningen University and Research Centre, Wageningen, The Netherlands

Online version at http://www.wur.nl/frontis

The titles published in this series are listed at the end of this volume

NEW APPROACHES TO THE ECONOMICS OF PLANT HEALTH

Edited by

ALFONS G.J.M. OUDE LANSINK

Business Economics, Wageningen University and Research Centre,
Wageningen, The Netherlands

 Springer

A C.I.P. Catalogue record for this book is available from the Library of Congress.

ISBN-10 1-4020-5825-X (HB)
ISBN-13 978-1-4020-5825-7 (HB)
ISBN-10 1-4020-5826-8 (PB)
ISBN-13 978-1-4020-5826-4 (PB)

Published by Springer,
P.O. Box 17, 3300 AA Dordrecht, The Netherlands.

www.springer.com

Printed on acid-free paper

CONTENTS

Methods for modelling non-monetary impacts of phytosanitary policies

Costs and benefits of phytosanitary measures

Economic and biophysical aspects of plant health policies

PREFACE

This book presents the outcomes of a workshop around the emerging area of the economics of plant health. The workshop was organized in Wageningen in July 2005 under the auspices of Frontis – Wageningen International Nucleus for Strategic Expertise.

Plant health nowadays plays an increasing role in national and international policy making. This explains the interest of the Netherlands Ministry of Agriculture, Nature and Food Quality in this workshop. The increasing importance of plant health in international policy making also follows from the recent establishment of a scientific panel on plant health by the European Food Safety Authority. This panel has to advise the EU on policy issues in the area of plant health. Plant health issues have numerous economic dimensions. Measures to control quarantine diseases and invasive species are usually costly, whereas the potential benefits, e.g., avoided losses, are often difficult to quantify. Quantifying the costs and benefits requires close collaboration between economists and epidemiologists. New GIS tools can play an important role in visualizing and modelling the combined economic and epidemiological consequences of control measures. Quarantine organisms and invasive species also frequently have impacts that go beyond agriculture. Impacts on landscapes and the environment call for the application of new approaches to measuring the economic impacts on society. This book presents a number of new approaches to economic modelling of plant health; it is primarily intended for policy makers and scientists working in the area of plant health. The methods presented here, however, have wider applicability in the economics of food safety and animal health.

ACKNOWLEDGEMENTS

I wish to express my gratitude to all authors for their excellent cooperation in preparing this book. Furthermore, this book could not have been produced without the help and advice of several people. Jan Schans is gratefully acknowledged for his comments on the first chapter. Rob Bogers (Frontis) has guided the editing process of the book; Paulien van Vredendaal (Wageningen UR Library) was responsible for the technical editing. Finally, Petra van Boetzelaer (Frontis) provided crucial assistance in the organization and logistics of the workshop.

Finally, I thank the Netherlands Ministry of Agriculture, Nature and Food Quality for kindly providing financial support for the organization of the workshop.

Alfons Oude Lansink,

Wageningen, July 2006

CHAPTER 1

NEW APPROACHES TO THE ECONOMICS OF PLANT HEALTH

General introduction

ALFONS G.J.M. OUDE LANSINK

*Business Economics, Wageningen University, P.O. Box 8130,
6700 EW Wageningen, The Netherlands*

INTRODUCTION

The protection of plant health and natural habitats is an important issue in many countries all over the world. Although the probability of a catastrophe due to introduction and spread of new plant diseases is generally very low, the economic and environmental impacts can be extremely large if a pest does occur (Knowler and Barbier 2000). OTA (1993) estimated the cumulative damage costs from alien species in the USA to be about $100 billion, for a selected group of pests over 85 years; for all introduced pests (some 50,000 species), the costs were estimated at $123 billion per year (Pimentel et al. 1999). Moreover, the probability of introducing non-indigenous organisms is rising due to: (1) increase in world trade; (2) population mobility and tourism; (3) increase in information, communication, technology and wealth in developed countries creating a demand for exotic plant material; (4) habitat fragmentation that may increase vulnerability; and (5) a general tendency towards the globalisation of the world economy (Shogren 2000).

In most countries, national plant health services have been assigned the task of safeguarding the national plant health situation. Under EU legislation, national plant health services are required to inhibit the introduction and spread of organisms harmful to plants or plant products (Plant Protection Service 2004). A distinction is made between harmful organisms not known to occur in the EU (A1 organisms), harmful organisms known to occur in the EU (A2 organisms), and harmful organisms that are not known in certain protected areas (B organisms). The plant health situation in particular is more generally affected by actions of all those actors

A.G.J.M. Oude Lansink (ed.), New Approaches to the Economics of Plant Health, 1–3.
© 2007 *Springer*.

involved in activities related to plant production and trade. The risk of a pest occurring has to be considered as an endogenous variable, influenced by economic activities that *create* risk and by economic activities that *reduce* risk. Examples of the former are international trade in plants and plant materials, production of crops, tourism and more general transport of goods and people (see, e.g. Dalmazzone 2000). Examples of activities aimed at reducing phytosanitary risks are: activities related to eradication, containment, prevention and monitoring of pests and diseases.

Phytosanitary policy worldwide is placed in the context of the WTO agreement on *Sanitary and Phytosanitary Measures* (SPS agreement); for alien species, policy is placed in the context of the Convention on Biological Diversity (CBD). Among others, the SPS agreement requires a scientific underpinning for trade-restricting measures based on the international standards, guidelines and recommendations developed under the auspices of the Secretariat of the IPPC. An example of such a standard is the standard for pest risk analysis in ISPM-11 (FAO 2003). More generally, governments worldwide are under increasing pressure of stakeholders from business and society. Stakeholders from business demand a rationalisation of phytosanitary policy in order to avoid unjustified trade restrictions and increase the efficiency of the plant production system. Stakeholders from society want to increase the scope of phytosanitary policy from the protection of the economic plant production system to the natural ecosystem.

SCOPE OF THE BOOK

Substantial methodological advances still have to be made regarding the economic modelling of harmful organisms as referred to in the SPS. The underlying book presents a number of methodological advancements on modelling and measuring the economic and environmental risk of harmful organisms. The book contains twelve papers that were presented at the Frontis Workshop on the Economics of Plant Health that was organised in Wageningen from 4 to 6 June 2005.

The book begins in Part I with two papers that aim at measuring the costs and benefits of phytosanitary measures in Finland and the UK, respectively. This is followed in Part II by a number of studies that evaluate the risks and economic effects of quarantine measures using spatially explicit bio-economic models. The studies in this part provide applications to Potato Brown Rot in the Netherlands and Asiatic Citrus Canker in the US. Next, Part III presents three novel attempts to model inspection policy regimes in the Netherlands and the US. The general problem addressed in these papers is: how to allocate scarce resources available for inspection of imported plants and plant materials so as to minimize the risks of pest incursion. Part IV presents two studies providing general frameworks for analysing the environmental and economic risks of pests and invasive species. The two chapters have a broad scope as they address the risks for the economic and ecological systems from harmful organisms and invasive species. Finally, Part V of the book demonstrates the usefulness of *Contingent Valuation Method* and *Multicriteria Analysis* in modelling the (non-monetary) impacts of plant health policies on the environment and society at large. These tools are useful in measuring impacts of plant health policies on natural habitats, landscapes and society at large.

ACKNOWLEDGEMENTS

The author thanks Jan Schans for helpful comments on an earlier version of this paper.

REFERENCES

Dalmazzone, S., 2000. Economic factors affecting vulnerability to biological invasions. *In:* Perrings, C., Williamson, M. and Dalmazzone, S. eds. *The economics of biological invasions*. Elgar, Cheltenham, 17-30.

FAO, 2003. *Pest risk analysis for quarantine pests, including analysis of environmental risks and living modified organisms*. Secretariat of the International Plant Protection Convention FAO, Rome. ISPM no. 11. [http://www.fao.org/DOCREP/006/Y4837E/Y4837E00.HTM]

Knowler, D. and Barbier, E.B., 2000. The economics of an invading species: a theoretical model and case study application. *In:* Perrings, C., Williamson, M. and Dalmazzone, S. eds. *The economics of biological invasions*. Elgar, Cheltenham, 70–93.

OTA, 1993. *Harmful non-indigenous species in the United States*. Office of Technology Assessment, United States Congress, Washington.

Pimentel, D., Lach, L., Zuniga, R., et al., 1999. *Environmental and economic costs associated with non-indigenous species in the United States*. College of Agriculture and Life Sciences, Cornell University, Ithaca. [http://www.news.cornell.edu/releases/Jan99/species_costs.html]

Plant Protection Service, 2004. *Summary of the plant health regulations of the Netherlands*. Ministry of Agriculture, Nature and Food Quality, Plant Protection Service, Wageningen. [http://www9.minlnv.nl/pls/portal30/docs/FOLDER/MINLNV/LNV/UITVOERING/UD_PD/ENGELS/PLANTHEALTHREG.PDF]

Shogren, J., 2000. Risk reduction strategies against the explosive invader. *In:* Perrings, C., Williamson, M. and Dalmazzone, S. eds. *The economics of biological invasions*. Elgar, Cheltenham, 56–69.

EFFICIENT INSPECTION POLICIES

CHAPTER 2

ROBUST INSPECTION FOR INVASIVE SPECIES WITH A LIMITED BUDGET

L. JOE MOFFITT, JOHN K. STRANLUND, BARRY C. FIELD AND CRAIG D. OSTEEN

Department of Resource Economics, University of Massachusetts, Amherst and U.S. Department of Agriculture, Economic Research Service.
E-mail: moffitt@resecon.umass.edu

Abstract. Invasive species can inflict significant costs on agriculture. Approaches to prevent introduction and/or to contain introduced species can also be very costly. Approaches to managing invasive-species problems include pre-emptive actions against potential invaders in foreign locales, border activities such as inspections to prevent introductions across international boundaries, domestic monitoring and control to prevent establishment if introductions occur, management of domestically established introductions through use of various forms of interference (e.g., vector control, enemies, pathogens, symbionts, endophytes, hosts, and/or physical factors perhaps as part of areawide management programs), and combinations of these approaches. This paper focuses on providing applicable quantitative decision support to the process of establishing efficient protocols for border protection under the severe uncertainty and resource constraints that characterize the inspection process. In this paper, a hybrid info-gap model is used in conjunction with stochastic dominance to develop a cost-effective protocol for invasive-species detection efforts. The model is illustrated by a detection problem faced at international ports. Problem characteristics advantageous to robust preparedness protocols are investigated.
Keywords: risk; stochastic dominance; severe uncertainty

INTRODUCTION

Productivity levels that have been achieved in modern economies through specialization and centralization may have been accompanied by vulnerabilities in the biosecurity realm that are only recently receiving attention from economists (Perrings et al. 2000). Modern agricultural, food, and health-care systems may present special challenges from a biosecurity perspective though other economic sectors such as transportation and education may also be based on bio-susceptible foundations (Wheelis et al.). Major resource commitments may be required to reduce both long- and short-term vulnerabilities in such systems.

Of particular concern in the United States are invasive species. An invasive species is defined as a species that does not naturally occur in a specific geographic

A.G.J.M. Oude Lansink (ed.), New Approaches to the Economics of Plant Health, 7–22.
© 2007 *Springer.*

area and whose introduction does or is likely to cause economic or environmental harm (Office of the President 1999). Included are species of plants, animals and other organisms such as microbes. By definition, a non-native species need not cause direct harm to be invasive. For example, the glassy-winged sharpshooter, recently introduced in California, causes relatively little harm by itself but vectors an important native plant pathogen.

As evidenced by weed pests such as purple loosestrife, which arrived via sailing ships in the 1800s, the presence of invasive species in the United States is not a new phenomenon. Moreover, as was the case with purple loosestrife, human actions have traditionally been the primary means of invasive species introductions. What may be different in more recent times is an increase in the frequency with which such exotic pests arrive. For example, so called 'jet age' insect pests such as Comstock mealybug, Egyptian alfalfa weevil, cereal leaf beetle, Russian wheat aphid, and pink hibiscus mealybug may all have come to areas of the United States aboard commercial aircraft (eg. Ervin et al. 1983; White et al. 1995; Moffitt et al. 1993; Moffitt 1999; Ogrodowczyk and Moffitt 2001). Moreover, there does not seem to be a shortage of potential invaders. One study estimates that there are 6,000 insect pests not in the United States but known to cause harm in foreign areas having ecological equivalents to the United States (McGregor 1973). A recent study takes an economics perspective in identifying a number of potentially important invaders (Moffitt and Osteen 2004).

Preparedness for introduction of invasive species may become much more important in the years ahead than it has been in the past. The trend toward reduction in barriers to trade may increase the volume of goods traded internationally and thereby increase the opportunity for introduction of biological materials across international boundaries. In addition, the development of the internet is rapidly increasing the volume of small-scale commerce in biological commodities globally and is adding greatly to the difficulty faced by regulatory authorities in protecting domestic environments. Finally, there is a growing awareness of the potential for related terrorist activities. The latter, in particular, adds to the uncertainty associated with biosecurity preparedness in an unprecedented way.

According to some recent estimates, the cost of improved biosecurity in the United States alone will be billions of dollars (Endress 2002; O'Hanlon et al. 2002). Internationally, it can only be presumed that costs will be perhaps prohibitive in many cases. In all cases, difficult management decisions will be made about the level of security that will be provided in order that efforts be sustainable. All of these decisions will be made under the severe uncertainty that characterizes biosecurity efforts generally.

This paper focuses on a new approach to developing a cost-effective strategy for managing biosecurity risk under severe uncertainty through detection effort. A hybrid info-gap model (Ben-Haim 2001b) is used in conjunction with stochastic dominance to determine an optimal robust strategy in invasive species detection. While the foundation of the model is explicitly info-gap in its use of a performance requirement, the nature of the performance requirement utilized here extends the hybrid info-gap approach to account for risk preference. Integration of stochastic dominance conditions into the hybrid info-gap framework facilitates traditional risk management under severe uncertainty.

The next section provides background for the economic model presented in the third section. The fourth section illustrates use of the economic model to determine an optimal robust strategy in a detection problem faced at a port of entry. The fifth section investigates the characteristics of the problem that are advantageous to use of robust preparedness protocols. Some concluding remarks are given in the final section.

RISK, UNCERTAINTY AND ROBUSTNESS

In the decades since Knight (1921) made a distinction between decision making under risk and decision making under uncertainty, most related economic research has focused on decision making under risk; that is, decision making with a probability distribution assigned to uncontrolled events (Hamouda and Rowley 1996). Traditional decision criteria under risk include mean-variance analysis and expected utility maximization with many applicable criteria having roots in the latter. Despite the economic research emphasis on risky decision making, there has often been difficulty in measuring and interpreting probability distributions associated with uncontrolled events and concern about risk assessment has been evident among researchers and practitioners alike. For example, in the preface of the 1935 edition of his book, Knight himself remarked "... I (am) still puzzled at the insistence of many writers on treating the uncertainty of result in choice as if it were a gamble on a known mathematical chance ...". In view of these concerns, several non-probabilistic alternatives for measuring risk have emerged (e.g. Katzner 1998; Ben-Haim 1999) and interest in alternatives among private- and public-sector managers is apparent.

Though receiving less emphasis than decision making under risk, decision making under uncertainty has not been ignored by economists. Traditional applied decision criteria under uncertainty include the maximin, maximax, Laplace, and Hurwitz criteria (see e.g. Render et al. 2003). While none of these criteria require knowledge of probability distributions for application, the first two represent polar extremes in terms of optimism and pessimism while the latter two require information similar to probabilities in order to be applied. Similarly, quantification of other notions related to uncertainty such as ignorance and surprise have also required specification of functions confined to the unit interval (Katzner 1998; Horan et al. 2002). Additionally, Kelsey (1993) developed a distinctive decision theory requiring a ranking of event probabilities rather than a specific probability distribution. Perhaps for these reasons, none of these decision criteria under uncertainty have achieved the widespread application in economics afforded traditional risk criteria.

Some recent decision theory research has focused on the notion of robustness in decision making as a means of coping with uncertainty. An important contribution due to Broens and Klein Haneveld (1995) not only formalizes the distinction between robustness and flexibility but also utilizes a notion of robustness that mirrors the most modern contributions to to this research area. Ben-Haim (1999) has developed and utilized a single-parameter characterization of uncertainty known as information (info)-gap decision theory (Ben-Haim 1994; 1999; 2001a). Info-gap decision theory is designed for decisions made under uncertainty; that is, for cases in which probability distributions for uncontrolled events are not available. The essence of info-gap is pursuit of a performance requirement over the largest possible 'range' of uncontrolled events. This concept of robustness is identical to that considered by Broens and Klein Haneveld (1995) in their analysis of natural gas investments which they refer to as commercial scope. A special case of the info-gap theory referred to as the hybrid model (Ben-Haim 2001b), treats a family of probability distributions as uncertain and seeks robustness with respect to the distributions. There have been a number of applications of the info-gap theory to problems ranging from selection of financial portfolios to optimal search in predator-prey systems (Ben-Haim 2001b).

The basic info-gap decision model due to Ben-Haim (1999) assumes that uncertainty about uncontrolled events cannot be characterized by probability and that realization of an event resolves uncertainty. The basic info-gap decision model is distinctive in utilizing a non-probabilistic characterization of uncertainty. In this decision-making environment, reward is a definite monetary amount that follows from a decision about a controlled factor and the realization of an uncontrolled event. A performance requirement, in terms of a monetary amount, is specified to guide decision making about the controlled factor. The robust optimal decision maximizes the 'range' of uncontrolled events over which the performance requirement is achieved. This notion of robustness is based analytically on a nested family of convex sets where the degree of nesting is characterized by a single parameter. In brief, the decision maker does not know the event faced; the basic info-gap model seeks a decision that is robust with respect to possible events.

The hybrid info-gap model (Ben-Haim 2001b) assumes that uncertainty about events can be characterized by a probability distribution but that the probability distribution is unknown. Uncertainty is resolved in this context by identification of the probability distribution rather than by realization of an event. In this decision-making environment, reward is a not a definite monetary amount but rather a probability distribution over rewards. Decisions need to be robust with respect to probability distributions rather than events.

In view of the nature of reward in the hybrid model, an extension of the performance requirement in the basic info-gap model is pursued here by specifying performance in terms of expected utility. The robust optimal decision in the extended hybrid model maximizes the 'range' of probability distributions over which the performance requirement is achieved while accounting for risk preference. Unlike traditional decision criteria under uncertainty, this extension of the hybrid info-gap model integrates traditional expected utility-based risk considerations into a decision model in which event probabilities are unknown. In brief, the decision maker does not know the gamble faced; the hybrid info-gap model developed here seeks a decision that is robust with respect to possible gambles while accounting explicitly for decision maker preferences for bearing risk.

The next section presents a hybrid decision model based on the info-gap theory with expected utility as a measure of performance. While information requirements of the performance measure may appear at first prohibitive, well known results from the economic theory of stochastic efficiency suggest otherwise. Stochastic dominance can be used to facilitate application of the model as is illustrated in modelling inspections for invasive species at international ports in section four.

THE MODEL

The following model depicts allocation of scarce resources to condition the probability density function of a random variable, taken to be reward, when the probability density function of reward is itself uncertain. Because of the uncertainty about the probability density function of reward, resources are allocated in order to achieve an outcome that meets a performance requirement and is as robust as possible with respect to the specification of the probability density function.

Let ε be a random variable, x be a vector of decision variables which impact the probability density function of ε, $f_{\varepsilon|x}$ be the probability density function of ε conditional on x, g be a probability density function for ε used in specifying a performance requirement, and $U(\varepsilon)$ be a von Neumann-Morgenstern utility function. With this notation, in the terminology of info-gap (Ben-Haim 2001b), the *system model* defines rewards and is taken to be expected utility, $\overline{U}_{(\cdot)}$, where the expectation is evaluated with respect to the subscripted probability density function. Note that in this context, the latter is not assumed to be known. The *uncertainty model* incorporates prior information in the system model and, in the case of the hybrid model, consists of a set of conditional probability density functions $\{ f_{\varepsilon|x} \}$. The *robustness function,* α (x), expresses the level of uncertainty over which the *performance requirement* (smallest acceptable reward) $\overline{U}_g$, will be achieved.

The *robust optimal decision* solves

$$\text{Maximize } \alpha(x) \qquad (1)$$
$$\scriptstyle (x)$$

$$\textit{Subject to } \overline{U}_{f_{\varepsilon|x}} \geq \overline{U}_g \qquad (2)$$

$$\int f_{\varepsilon|x}\, d\varepsilon = 1 \qquad (3)$$

$$f_{\varepsilon|x} \geq 0 \qquad (4)$$

$$x \in X \qquad (5)$$

where the set X reflects any constraints on x other than the performance requirement and those constraints on the conditional density $f_{\varepsilon|x}$ related to the definition of a probability density function. Assuming a solution exists, the solution to (1) - (5) provides a specific value of the vector of decision variables, x^*, and associated conditional density function $f_{\varepsilon|x^*}$. The latter is superior to the performance requirement in terms of expected utility and maximizes robustness. Given an appropriate specification of robustness, the latter condition suggests that the performance requirement will be achieved not only under $f_{\varepsilon|x^*}$ but also under perhaps a wide range of related densities.

As it stands, the model (1) - (5) poses a very difficult constrained optimization problem mainly because its information requirements seem so extensive. Two key elements needed to implement (1) - (5) include specification of the robustness objective function and the performance requirement which is shown in (2) as a constraint on expected utility. As demonstrated in the next section, it is possible to make meaningful specifications for both of these elements.

The robustness function, $\alpha(x)$, reflects the conditions under which the performance requirement will be achieved and can be specified in different ways (Ben-Haim 2001b). In the hybrid info-gap model, the elements in the uncertainty model are probability density functions. An intuitive interpretation of robustness, consistent with the spirit of its usage in the basic info-gap model, suggests that at the optimal solution to (1) - (5), not only does $f_{\varepsilon|x^*}$ achieve the performance requirement but other related conditional densities, one of which may turn out to be the correct one, do likewise. A specification of the robustness function suitable for all cases is perhaps not possible or even necessary. A number of criteria including Euclidian distance, variance, relative entropy, Gini's mean difference and a host of other measures can be utilized to identify, in a particular sense, the least advantageous probability density function that achieves the performance requirement in order to maximize the potential for achieving the performance requirement under a wide "range" of densities (Ben-Haim 2001b; Ebrahimi et al. 1999; Hansen and Sargent 2001; Yitzhaki 1982). In the case where the different conditional densities are characterized by a single parameter, Euclidian distance provides an intuitive measure of robustness. For the model presented in the next section, both variance and entropy will also provide equivalent measures.

If the expected utility of $f_{\varepsilon|x}$ exceeds the expected utility of g; i.e., $\overline{U}_{f_{\varepsilon|x}} = \int_{-\infty}^{\infty}$

$U(\varepsilon) f_{\varepsilon|x} \, d\varepsilon > \int_{-\infty}^{\infty} U(\varepsilon) g d\varepsilon = \overline{U}_g$, then $f_{\varepsilon|x}$ is preferred to g as required by (2). An important result in the economic theory of stochastic efficiency based on expected utility is known as second-degree stochastic dominance (SSD) (Mas-Colell et al. 1995). In brief, SSD can be stated as follows: a risk-averse individual will prefer $f_{\varepsilon|x}$ to g if and only if $\int_{-\infty}^{\varepsilon} (G(t) - F_{\varepsilon|x}(t)) dt \geq 0$ for all ε with a strict inequality for at least one ε where $F_{\varepsilon|x}$ and G denote the cumulative distribution functions associated with $f_{\varepsilon|x}$ and g, respectively. The significance of the SSD criterion is that it permits comparison of different gambles over the class of risk-averse individuals using only the cumulative distribution functions of the gambles; i.e., individual utility functions need not be known in order to implement the constrained optimization of robustness depicted in (1) - (5). Figure 1 depicts the SSD conditions diagrammatically for the case where the cumulative distribution functions cross only once. Note that in the figure, SSD requires only that the area labelled 'A' exceed the area labelled 'B'. In cases in which the cumulative distribution functions cross more than once, such as the illustration presented in the next section, similar graphical conditions can be identified.

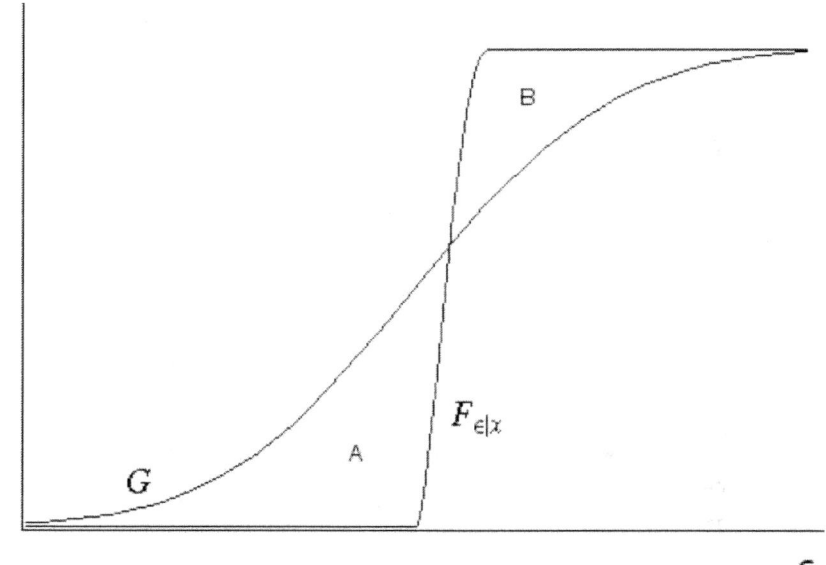

Figure 1. $f_{\varepsilon|x}$ *dominates g by SSD since area A is greater than area B*

ROBUST DETECTION AT AN INTERNATIONAL PORT OF ENTRY

This section illustrates use of the model developed in the previous section for allocating scarce resources to manage invasive-species risk under uncertainty. Risks are managed through detection effort consisting of inspection of shipments for the presence of invasive species. Hence, use here of the model developed in the previous section is intended to focus on a robust detection effort.

Let B denote the benefit due to shipping activity at a port of entry without an invasive-species threat, p denote the probability that an invasive species is present on one of N shipments that will call at the port, L denote the cost of failure to prevent passage of the species through the port, and n denote the number of shipments inspected at cost c(n) where $c' > 0$, $c'' > 0$, and c(0) = f. Hence, the cost function is assumed to be increasing at an increasing rate; that is, cost is a strictly convex function of number of shipments inspected. Fixed costs, f, are permitted but may be zero. Budgetary considerations limit expenditure on detection effort to C_0. The probability, p, is completely unknown; hence, a hybrid info-gap model is used to investigate optimal inspection.

The conditional probability density function, $f_{\varepsilon|x}$, for net benefit, ε, due to activity at the port where x = (n, p) is

$$f_{\varepsilon|x}(\varepsilon) = \begin{array}{l} 1 - \frac{p(N-n)}{N}; \text{if } \varepsilon = B - c(n) \\ \frac{p(N-n)}{N}; \text{if } \varepsilon = B - L - c(n) \end{array} \tag{6}$$

Note that a probability density function for net benefit is associated with each possible p and each possible value of detection effort, n.

Gauging the performance of stochastic systems by the probability of failure and establishing a performance requirement in terms of failure probability are common. To establish a performance requirement for detection effort in terms of the probability of biological material passing through the port undetected (failure probability), note again that this probability is given by $\frac{p(N-n)}{N}$. Let p_c denote the largest acceptable value for this probability. Note that p_c is realized as the actual probability of biological material passing through the port undetected if and only if p = 1 and n = N(1 - p_c). As it stands, p_c is a performance parameter without any economic component. However, the unique probability density function over net benefit associated with p_c makes evident its economic implications. Call the associated probability density function g where

$$g(\varepsilon) = \begin{array}{l} 1 - p_c \ ; \text{if } \varepsilon = B - c(N(1 - p_c)) \\ p_c ; \text{if } \varepsilon = B - L - c(N(1 - p_c)) \end{array} \tag{7}$$

According to (7), any inspection errors including failure to detect as well as fall alarms, are assumed to be negligible. With this definition of g, $\bar{U}g$ in the model (1) - (5) is $\bar{U}g$ = U(B - c(N(1 - p_c))) (1 - p_c) + U(B - L - c(N(1- p_c))) p_c . The performance requirement, $\overline{U}_{f_{\varepsilon|x}} \geq \bar{U}g$ in (1) - (5) is then

$$\text{U(B - c(n))}(1 - \frac{p(N - n)}{N}) + \text{U(B - L - c(n))}\frac{p(N - n)}{N} \geq \tag{8}$$
$$\text{U(B - }c(N(1 - p_c)))(1 - p_c) + \text{U(B - L - }c(N(1 - p_c)))p_c$$

Note that under the assumptions about U, second-degree stochastic dominance can be used to express (8) in simpler terms that do not involve an expression for the utility function. This is accomplished by comparing cumulative distribution functions according to SSD conditions as depicted in Figure 1. The cumulative distribution functions in this case are step functions corresponding to the discrete probability density functions specified for $f_{\varepsilon|x}$ and g. In particular, SSD conditions reveal that (8) holds if and only if p_c ((B - L - c(n)) - (B - L - c(N(1 - p_c)))) $\geq$ ($\frac{p(N-n)}{N}$ - p_c) ((B - c(N(1 - p_c)) - (B - L - c(n))). The preceding inequality simplifies to become p_c L - $\frac{p(N-n)}{N}$ (L + c(n) - c(N(1 - p_c))) $\geq$ 0. Given expressions (6) and (7) for $f_{\varepsilon|x}$ and g, respectively, the latter inequality corresponds to a relationship between areas under cumulative distribution functions similar to those areas depicted in Figure 1 as areas A and B.

Since the uncertainty model consists of a set of probability density functions characterized by a single parameter (p) confined to the unit interval, the robustness function can be specified meaningfully as that parameter. Maximizing robustness in this case means selecting detection effort to identify the largest value of p for which the model constraints hold. The implication of the robust optimal decision is that budgetary and performance requirements will be achieved for smaller values of p as well.

In this case, the model (1) - (5) is

$$\text{Maximize} \atop (n, p) \tag{9}$$

$$\text{Subject to } \frac{p(N-n)}{N} \text{ (L + c(n)- c(N(1 - } p_c \text{))) - } p_c \text{ L } \leq 0 \tag{10}$$

$$p \leq 1 \tag{11}$$

$$c(n) \leq C_0 \tag{12}$$

$$n, p \geq 0 \tag{13}$$

Characteristics of this problem that are expected to offer advantages to risk averse decision makers for use of the model (9) - (13) are investigated in the next section.

Use of (9) - (13) is illustrated by considering detection effort at a hypothetical, moderate-sized container port. Consider a port at which 1,000 vessels call annually handling cargo with an estimated value of $25 billion. The port generates tax revenue along with jobs providing a significant sum of wages each year. A biosecurity failure that interrupts activity at the port for an extended period of time will cost approximately $100 million annually. The port commission has recently implemented heightened security measures in response to the threat of invasive species. The annual budget reveals $5 million is allocated for inspection. Cost for complete inspection of n vessels is estimated to be $c(n) = 1000 + 100 \ n^2 - 100 \ n$. Assuming a failure probability P_c = 0.05, complete inspection of $N(1- P_c) = 950$ vessels costs over $90 million per year and is not possible within the port's security budget, which permits complete inspection of only 224 vessels each year. Solving (9) - (13) reveals n^* = 68 vessels and p^* = 0.521. Annual inspection cost for 68 vessels is approximately $5 million. The cumulative distribution functions associated with robust optimal detection ($F_{\varepsilon|x^*}$) and the failure probability (G) are

shown in Figure 2. For any p < 0.521, $F_{\varepsilon|x^*}$ is preferred to G by all risk-averse

decision makers. Moreover, spending the entire security budget to inspect 224 vessels provides a probability distribution which is less robust (preferred by risk averse decision makers for a smaller range of values for p) than that achieved by inspecting only 68 vessels.

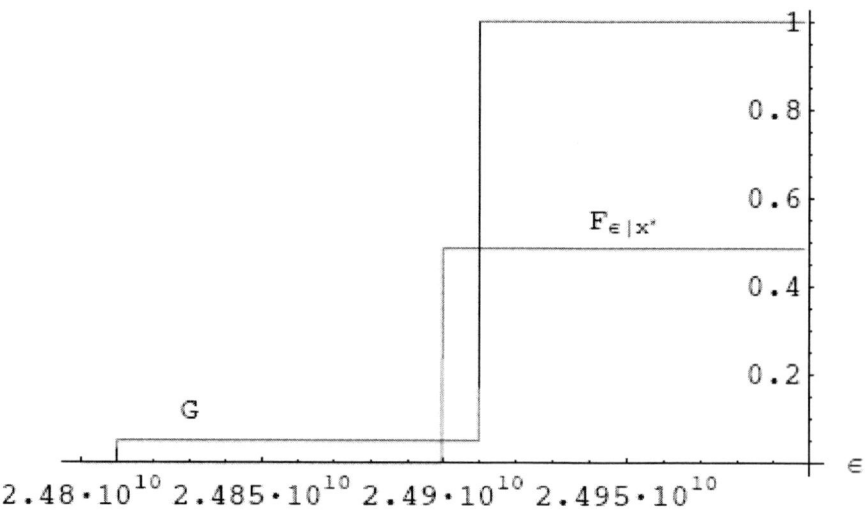

Figure 2. Cumulative distribution functions associated with inspection of 950 vessels (G) and 68 vessels ($F_{\varepsilon|x^*}$)

This illustration enables some additional insight into the nature of a robust protocol based on the hybrid info-gap model with performance expressed in terms of a risk averter's expected utility. The outcome of the robust protocol (spend about half of the budget to inspect only 68 vessels rather than spend the entire budget to inspect 224 vessels) seems perhaps counterintuitive at first glance. At a purely intuitive level, perhaps many would expect to see robustness attached to risk-averse decision makers inspecting a larger rather than a smaller number of vessels. On the other hand, rationalizing the optimality of the lower detection effort as due simply to a convex cost function abstracts away from the essential fact that the detection problem involves decision making under uncertainty and decision-maker risk preferences. Some additional insight into the optimal robust inspection effort can be obtained by focusing on the relationship between maximum robustness and inspection level.

To see more clearly why inspecting only 68 vessels subject to available resources is preferred by all risk-averse decision makers for a larger range of values of p than inspecting 224 vessels, consider Figure 3. The figure displays the largest value of p at each inspection level, n, for which the resulting pdf of reward is preferred by all risk averters to the pdf associated with the failure probability. Note from the figure, that if budgetary considerations restrict inspections to less than approximately 900 vessels, then the largest value of p is found where n is 68. The inspection effort associated with this relative internal maximum in Figure 3 is the n^* associated with $F_{\varepsilon|x^*}$ in Figure 2.

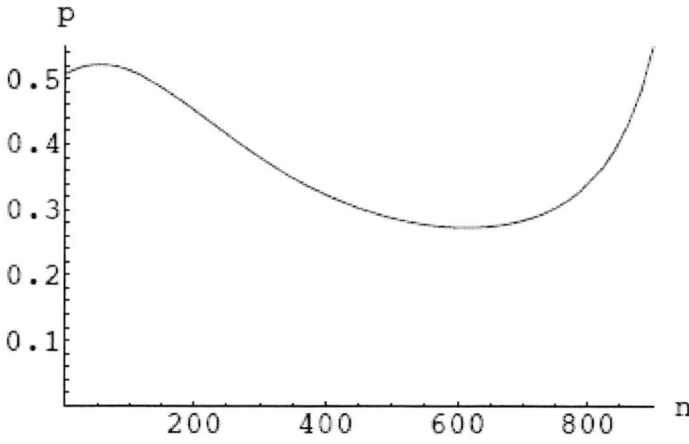

Figure 3. Largest p given n for which a risk averter prefers the gamble based on n to the gamble based on the failure probability, p_c

Identification of the factors that contribute to the shape of the relationship between robustness and inspection level helps to discern the source of the advantages that may be forthcoming from use of robust protocols in the presence of a resource constraint. Consider the shape of the relationship depicted in Figure 3 as inspection increases from n = 0. As n rises, both the 'downside' risk, $\frac{p(N-n)}{N}$, and reward, B - L - c(n), associated with $F_{\varepsilon|x}$ are affected. The impact on 'downside' risk initially dominates the impact on reward permitting greater robustness. This forms the initial upper sloping portion of the robustness-inspection level relationship shown in Figure 3. This relationship between risk and reward impacts persists until a relative internal maximum for robustness occurs at n = 68. Beyond this point, the convexity of the cost function yields reward impacts to further inspection that dominate, at least for a range of inspection levels, the associated risk impact. As a result, robustness declines steadily until approximately n = 625, where robustness reaches a relative internal minimum. As inspection effort grows larger still, the hyperbolic relationship between p and n in the expression for 'downside' risk assumes primacy enabling robustness to rise again despite the convexity of inspection cost and its impact on reward. Hence, without resource constraints, G will prevail; with even a modest constraint, a much different inspection level is preferred.

The relative internal maximum represented by $F_{\varepsilon|x^*}$ reflects the most robust balance between risk and reward relative to the balance embodied in G that is sufficient for any risk averter with a limited budget. In this illustration, this result follows from the nature of inspection cost and inspection's impact on downside risk. The next section investigates in more detail the characteristics of this problem that enable robust protocols to offer advantages over simpler alternatives for risk-averse decision makers.

CHARACTERISTICS AND POTENTIAL ADVANTAGE FOR ROBUST PROTOCOLS

This section considers, in more detail, characteristics of the problem considered in the previous section that make use of a robust protocol potentially advantageous for risk-averse decision makers. In brief, the objective of this section is to determine problem characteristics that enable rapid screening for potentially fruitful application of (9) - (13). Two propositions are used to assist in identification of relevant characteristics. Proposition 1 helps to define the nature of the solution to (9) - (13) that holds potential for a robust protocol to represent an improvement over standard protocols such as expenditure of a budgeted amount or pursuit of a target failure probability. Proposition 2 extends the result of Proposition 1 to more basic and readily recognizable problem characteristics.

At an intuitive level, if the constraint on robustness (11) is binding or if the budget constraint (12) is binding, resource allocation under a robust protocol may not differ much from simpler alternative protocols. In such cases, either resources permit maximum possible robustness to be achieved or else robustness is maximized by expenditure of all available resources, respectively. In neither case is there potential for more cost-effective risk management through detection efforts. The following proposition confirms this reasoning.

Proposition 1: The robust optimal decision associated with (9) - (13) offers a more desirable probability distribution for risk-averse decision makers than would be achieved through expenditure of a budgeted amount or pursuit of a target failure probability if and only if the performance requirement (10) is the sole binding constraint; i.e., $n^* < \min(c^{-1}(C_0), N(1-P_c))$ if and only if $\frac{p^*(N-n^*)}{N}(L + c(n^*) - c(N(1 - P_c))) - P_c L = 0$, $n^* > 0$, $p^* > 0$, $p^* < 1$, and $c(n^*) < C_0$.

Proof: The Kuhn-Tucker Lagrangian, L , and conditions associated with (9) - (13) are

$$L(n, p, \lambda_1, \lambda_2, \lambda_3) = p - \lambda_1 \left(\frac{p(N-n)}{N} (L + c(n) - c(N(1 - p_c))) - p_c L \right)$$
$$- \lambda_2 (p - 1) - \lambda_3 (c(n) - C_0) \tag{14}$$

$$\frac{\partial L}{\partial n} = \frac{p\lambda_1 (L + c(n) - c(N(1 - p_c))) + (np\lambda_1 - N(p\lambda_1 + \lambda_3))c'(n)}{N} \leq 0 \tag{15}$$

$$\frac{\partial L}{\partial p} = 1 - \lambda_2 - \frac{(N - n)\lambda_1 (L + c(n) - c(N(1 - p_c)))}{N} \leq 0 \tag{16}$$

$$\frac{\partial L}{\partial \lambda_1} = Lp_c - \frac{(N - n)p(L + c(n) - c(N(1 - p_c)))}{N} \geq 0 \tag{17}$$

$$\frac{\partial L}{\partial \lambda_2} = 1 - p \geq 0 \tag{18}$$

$$\frac{\partial L}{\partial \lambda_3} = C_0 - c(n) \geq 0 \tag{19}$$

$$n \frac{\partial L}{\partial n} = p \frac{\partial L}{\partial p} = \lambda_1 \frac{\partial L}{\partial \lambda_1} = \lambda_2 \frac{\partial L}{\partial \lambda_2} = \lambda_3 \frac{\partial L}{\partial \lambda_3} = 0 \tag{20}$$

$$n, p, \lambda_1, \lambda_2, \lambda_3 \geq 0 \tag{21}$$

Under regularity conditions, (14) - (21) characterize the solution to (9) - (13). If no subset of conditions (17) - (19) hold as equalities, then (15) and (16) admit no solution. Hence, a subset of conditions (17) - (19) must hold as equalities at the solution. If (18) holds as an equality, then $n^* = N(1 - p_c)$ while if (19) holds as an equality, then $n^* = c^{-1}(C_0)$. If (18) and (19) both hold as equalities, then $n^* = c^{-1}(C_0) = N(1 - p_c)$. If (10) is the sole binding constraint then (17) holds alone as an equality and $n^* < \min(c^{-1}(C_0), N(1 - p_c))$. On the other hand, if $n^* < \min(c^{-1}(C_0), N(1 - p_c))$ then neither (18) nor (19) can hold implying (17) holds alone as an equality and (10) is the sole binding constraint.

The implications of Proposition 1 can be made more transparent by considering conditions under which (17) holds as an equality at the solution to (9) - (13).

Proposition 2 provides the desired result.

Proposition 2: A robust protocol is potentially advantageous in detection problems for which the failure cost does not exceed inordinately the cost of complete inspection of the proportion of the population of vessels associated with the failure probability, i.e., L $\gg$ c(N(1 - p_c)).

Proof: Expressing (17) as an equality, solving for p, and maximizing the result as a function of n gives the following necessary and sufficient conditions:

$$L - c\big(N\big(1 - p_c\big)\big) = (N - n)c'(n) - c(n) \tag{22}$$

$$c'(n) < (1/2)(N - n)\, c''(n) \tag{23}$$

In view of Proposition 1, unless (22) and (23) are solvable for some n, then a robust protocol cannot be advantageous. According to (22), marginal 'failure' cost of incomplete inspection at N(1 - p_c) equals marginal 'physical' cost of incomplete inspection at n. According to (23), marginal physical inspection cost at n must be rising more slowly than marginal physical cost of inspecting the rest of the population of vessels. If the left-hand side of (22) is inordinately large, it will exceed the right-hand side and prevent (17) from holding as an equality ruling out a potential advantage for a robust protocol.

Assuming p_c is small, Proposition 2 suggests that problems for which the cost of complete inspection of a large proportion of the population of vessels does not differ inordinately from the failure cost, are those problems where pursuing the robust optimal decision may be beneficial for risk-averse decision makers. The result of Proposition 2 suggests that problems with catastrophic economic consequences relative to complete inspection costs for a large proportion of vessels cannot be pursued fruitfully with a robust protocol. In such cases, use of a failure probability approach or simply expending a budgeted amount will correspond to the solution to (9) - (13). In other cases, use of (9) - (13) may yield benefits as suggested by the numerical illustration in the preceding section.

CONCLUDING REMARKS

Current invasive-species inspection policy in the United States is based on sampling intensity to meet a detection probability requirement given the size of the population to be sampled. Such policy does not incorporate economic criteria, which may be increasingly important as inspection techniques grow more sophisticated and costly. Moreover, existing criteria may be infeasible as the population of shipments grows with increasing trade due to limited inspection resources.

Robust satisfying as exemplified by the info-gap theory may provide a feasible alternative to current inspection policy that recognizes both resource constraints and inherent uncertainty in the inspection process. The method described in this chapter provides for maximum robustness and, in the illustration given here, does so at a fraction of the cost of a non-economic inspection policy. Of particular interest is the realization that the information requirements associated with application of the info-gap approach are much less than those associated with a more traditional risk assessment and management approach. Avoiding the need for what may be costly risk assessment may well lower the bar for required information for policy by a non-trivial amount. Of course, if risks are known, then the traditional risk management framework may be used to advantage.

It is important to remain mindful that, in practice, decisions related to detection of invasive species contained in trade shipments are characterized by both important consequences and severe uncertainty. The hybrid info-gap model, extended to incorporate risk considerations for purposes of performance, provides an opportunity to manage both risk and uncertainty through robustness. Efficiency gains from robust protocols depend on problem characteristics and will be appropriate on a case-by-case basis. Extensions of the info-gap model described here to incorporate inspection errors such as failures to detect and false alarms are possible.

NOTES

Funding support provided by the USDA/ERS/PREISM Cooperative Agreement No. 43-3AEM-4-80115 and the Massachusetts Agricultural Experiment Station under project number MAS00861 is gratefully acknowledged. The views expressed are those of the authors and not necessarily those of the Economic Research Service or the U.S. Department of Agriculture.

REFERENCES

Ben-Haim, Y., 1994. Convex-models of uncertainty: applications and implications. *Erkenntnis*, 41 (2), 139-156.
Ben-Haim, Y., 1999. Set-models of information-gap uncertainty: axioms and an inference scheme. *Journal of The Franklin Institute*, 336 (7), 1093-1117.
Ben-Haim, Y., 2001a. Decision trade-offs under severe info-gap uncertainty. *In: 2nd International symposium on imprecise probabilities and their applications, Ithaca, NY, 26-29 June 2001.* [http://ippserv.rug.ac.be/~isipta01/proceedings/s001.pdf]
Ben-Haim, Y., 2001b. *Information-gap decision theory: decisions under severe uncertainty*. Academic Press.
Broens, D.F. and Klein Haneveld, W.K., 1995. Investment evaluation based on the commercial scope the production of natural gas. *Annals of Operations Research*, 59 (1), 195-226.
Ebrahimi, N., Maasoumi, E. and Soofi, E.S., 1999. Ordering univariate distributions by entropy and variance. *Journal of Econometrics*, 90 (2), 317-336.
Endress, L.H., 2002. *Terrorism and the economics of biological invasions*. Paper presented at the 77th annual Western Agricultural Economic Association International Conference, Seattle, WA, 2002.
Ervin, R.T., Moffitt, L.J. and Meyerdirk, D.E., 1983. Comstock mealybug: cost analysis of a biological control program in California. *Journal of Economic Entomology*, 76 (3), 605-609.
Hamouda, O.F. and Rowley, R., 1996. *Probability in economics*. Routledge, London. Frontiers of Political Economy no. 5.
Hansen, L.P. and Sargent, T.J., 2001. Robust control and model uncertainty. *American Economic Review*, 91 (2), 60-66.
Horan, R.D., Perrings, C., Lupi, F., et al., 2002. Biological pollution prevention strategies under ignorance: the case of invasive species. *American Journal of Agricultural Economics*, 84 (5), 1303-1310.

Katzner, D.W., 1998. *Time, ignorance, and uncertainty in economic models.* University of Michigan Press, Ann Arbor.

Kelsey, D., 1993. Choice under partial uncertainty. *International Economic Review,* 34 (2), 297-308.

Knight, F.H., 1921. *Risk, uncertainty, and profit.* Houghton Mifflin, Boston.

Mas-Colell, A., Whinston, M.D. and Green, J.R., 1995. *Microeconomic theory.* Oxford University Press, New York.

McGregor, R.C., 1973. *The emigrant pests: a report to the Animal and Plant Health Inspection Service.* Import Inspection Task Force.

Moffitt, L.J., 1999. *Economic risk to United States agriculture of pink Hibiscus mealybug invasion: a report to the US Department of Agriculture, Animal and Plant Health Inspection Service.* USDA-APHIS, Beltsville.

Moffitt, L.J., Allen, P.G. and Wu, F., 1993. *Benefit/cost analysis of biological control of the cereal leaf beetle: a report to the US Department of Agriculture, Animal and Plant Health Inspection Service.* USDA-APHIS, Beltsville.

Moffitt, L.J. and Osteen, C.D., 2004. *Cost criteria in crop protection.* A report to the US Department of Agriculture, Economic Research Service.

O'Hanlon, M.E., Orszag, P.R., Daalder, I.H., et al., 2002. *Protecting the American homeland: a preliminary analysis.* Brookings Institution Press, Washington.

Office of the President, 1999. *Invasive species.* Executive order 13112 of February 3, 1999, signed by President William J. Clinton. [http://ceq.eh.doe.gov/nepa/regs/eos/eo13112.html]

Ogrodowczyk, J.D. and Moffitt, L.J., 2001. *Economic impact of purple loosestrife in the United States: a report to the National Biological Control Institute prepared under Grant Number 00-8100-0542-GR.* University of Massachusetts, Amherst.

Perrings, C., Williamson, M. and Dalmazzone, S. (eds.), 2000. *The economics of biological invasions.* Elgar, Cheltenham.

Render, B., Stair, R.M. and Hanna, M.E., 2003. *Quantitative analysis for management.* 8th edn. Prentice Hall, Englewood Cliffs.

Wheelis, M., Casagrande, R. and Madden, L., Biological attack on agriculture: low-tech, high-impact bioterrorism. *BioScience,* 52 (7), 569-576.

White, J.M., Allen, P.G., Moffitt, L.J., et al., 1995. Economics of an areawide program for biological control of the alfalfa weevil. *American Journal of Alternative Agriculture,* 10 (4), 173-179.

Yitzhaki, S., 1982. Stochastic dominance, mean variance, and Gini's mean difference. *American Economic Review,* 72 (1), 178-185.

CHAPTER 3

ON ECONOMIC-COST MINIMIZATION VERSUS BIOLOGICAL-INVASION DAMAGE CONTROL

GREGORY DEANGELO[#], AMITRAJEET A. BATABYAL[##,1] AND SESHAVADHANI KUMAR[###]

[#] *Department of Economics, University of California at Santa Barbara, 2127 North Hall, Santa Barbara, CA 93106-9210, USA. E-mail: deangelo@econ.ucsb.edu*
[##] *Department of Economics, Rochester Institute of Technology, 92 Lomb Memorial Drive, Rochester, NY 14623-5604, USA. E-mail: aabgsh@rit.edu*
[###] *Department of Mathematics and Statistics, Rochester Institute of Technology, 85 Lomb Memorial Drive, Rochester, NY 14623-5603, USA. E-mail: sxksma@rit.edu*

Abstract. Recently, Batabyal et al. (2005) have used a queuing model to show that there is a *tension* between economic-cost minimization and inspection stringency in invasive-species management in the following sense: greater (lesser) inspection stringency with a larger (smaller) number of inspectors leads to higher (lower) economic costs. We use a *more general* queuing model to investigate whether there is, in fact, a tension between cost minimization and inspection stringency. Our theoretical analysis shows that there is no definite answer to this question. Therefore, we use numerical methods, and our numerical analysis leads to two conclusions. For many values of the model parameters that delineate the strictness of inspections, there *is* a tension between cost minimization and inspection stringency. In contrast, for most values of the model parameter that describes the volume of maritime trade handled by the port under study, there is *no* tension between cost minimization and inspection stringency.
Keywords: inspection; invasive species; maritime trade; queuing theory; uncertainty

INTRODUCTION

It is now common knowledge that maritime trade in goods comprises a substantial fraction of the world's total international trade in goods. Ships are the basic vehicle in maritime trade and therefore they are frequently used to carry all manner of goods in containers from one part of the world to another. International-trade theorists have shown that there are clear gains to the parties involved in such voluntary trade between the different nations of the world. Even so, with the passage of time, analysts have argued that the magnitude of these gains is likely to be less than what most researchers have hitherto believed. Why might this be the case? As Parker et al. (1999), Perrings et al. (2000a) and Batabyal (2004) have pointed out, this is

A.G.J.M. Oude Lansink (ed.), New Approaches to the Economics of Plant Health, 23–37.
© 2007 *Springer*.

because in addition to transporting goods in containers between nations, ships have also managed to transport a variety of non-native plant and animal species (also known as alien or invasive species) from one geographical part of the world to another.

There are two main ways in which ships have transported invasive species from one part of the world to another. First, many marine non-native species have been introduced into a nation, often unwittingly, by ships discarding their ballast water. Cargo ships usually carry ballast water in order to increase vessel stability when they are not carrying full loads. When these ships come into port, this ballast water must be jettisoned before cargo can be loaded. This manner of species introductions is significant and, very recently, the problem of managing invasive species that have been introduced into a particular nation by discarding ballast water has received some attention in the economics literature[2].

The second way in which invasive species have been introduced into a particular nation is by means of the containers that ships frequently use to carry cargo from one country to another. In this regard, the reader should note that non-native species can remain hidden in containers for extended periods of time. In addition, material such as wood – that is often used to pack the cargo in the containers – may itself contain invasive species. In fact, as noted by Batabyal and Nijkamp (2005), a joint report from the United States Department of Agriculture (USDA), the Animal and Plant Health Inspection Service (APHIS) and the United States Forest Service (USFS) has pointed out that nearly 51.8 % of maritime shipments contain solid wood packing materials and that infection rates for solid wood packing materials are non-trivial (USDA-APHIS 2000, p. 25). For example, inspections of wooden spools from China showed infection rates between 22% and 24 % and inspections of braces for granite blocks imported into Canada were found to hold live insects 32 % of the time (USDA-APHIS 2000, p. 27-28).

Non-native species are of interest to both economists and biologists because when these species invade new habitats, they impose tremendous costs on the nations in which these new habitats are located. To see this, consider the following two estimates of the magnitude of the economic costs for one country, namely, the United States. First, the Office of Technology Assessment (OTA 1993) has determined that the Russian wheat aphid caused US$ 600 million worth of crop damage between 1987 and 1989. Second, Pimentel et al. (2000) have approximated the total costs of all non-native species at around $ 137 billion per year.

In addition to the economic costs that we have just mentioned, alien species have caused considerable biological damage as well. In this regard, Vitousek et al. (1996) have explained that alien species can change ecosystem processes, act as vectors of diseases, and diminish biological diversity. Further, Cox (1993) has observed that out of 256 vertebrate extinctions with a known cause, 109 are the outcome of biological invasions. The discussion in this and the preceding paragraph together tell us that invasive species have been and continue to be a great menace to society.

It is only very recently that economists have acknowledged the consequences of the problem of biological invasions. As a result, Perrings et al. (2000b, p. 11) have rightly pointed out that "the economics of the problem has...attracted little attention". An implication of this regrettable state of affairs is that our knowledge of the economic and the management aspects of invasive species is deficient. Now, from the perspective of a manager, there are a number of actions that this individual can take to address the problem of biological invasions. It is helpful to divide these actions into pre-invasion and post-invasion actions. The objective of pre-invasion or prophylactic actions is to prevent non-native species from invading a new habitat. In contrast, post-invasion actions involve the optimal regulation of a non-native species, given that this species has already invaded a new habitat.

The small economics literature on biological invasions has, for the most part, focused its attention on the properties of alternate *post-invasion* actions. For instance, Barbier (2001) has pointed out that the economic effect of a biological invasion can be ascertained by studying the nature of the interaction between the native and the non-native species. He notes that the economic effect depends on whether this interaction involves interspecific competition or dispersion. Second, Eiswerth and Johnson (2002) have examined an intertemporal model of invasive-species stock management. These researchers note that the optimal level of management effort is responsive to ecological factors that are not only species- and site-specific but also stochastic in nature. Third, Olson and Roy (2002) have used a probabilistic framework to analyse the circumstances under which it is optimal to wipe out an invasive species and the circumstances under which it is not optimal to do so. Finally, Eiswerth and Van Kooten (2002) have demonstrated that in some situations, it is possible to use information supplied by specialists to develop a model in which it is optimal not to wipe out but instead regulate the spread of an alien species.

The regulation of a potentially detrimental non-native species *before* it has invaded a new habitat has been analysed by Horan et al. (2002), Batabyal et al. (2005) and Batabyal and Beladi (2006). Horan et al. (2002) examine the attributes of management approaches under full information and under uncertainty. Batabyal and Beladi (2006) study optimization problems stemming from the stationary-state analysis of two multi-person inspection regimes. Finally, Batabyal et al. (2005) note that there is a *tension* between economic-cost minimization and inspection stringency in invasive-species management in the following sense: greater inspection stringency with a larger number of inspectors leads to higher economic costs, and smaller inspection stringency with a smaller number of inspectors results in lower economic costs. The reader should understand that greater (smaller) inspection stringency reflects a heightened (diminished) concern for the potential damage from one or more biological invasions. Therefore, a port manager who places a relatively large (small) weight on invasion damage control will, *ceteris paribus*, want to inspect ships more (less) stringently.

Given the importance of the inspection function in invasive species management, the purpose of this paper is to investigate the generality of the 'tension result' in the Batabyal et al. (2005) paper. To undertake this investigation, we use a queuing model that is *more general* than the model used in Batabyal et al. (2005). Our theoretical analysis shows that there is *no* definite answer to the question as to whether there is or isn't a tension between economic-cost minimization and inspection stringency. Therefore, we use numerical methods, and our numerical analysis leads to two conclusions. First, for many values of the model parameters that delineate the strictness of inspections, there *is* a tension between economic-cost minimization and inspection stringency. Second and in contrast, for most values of the model parameter that describes the volume of maritime trade handled by the port under study, there is *no* tension between economic-cost minimization and inspection stringency.

The rest of this paper is organized as follows. The next section provides a conceptual framework based on queuing theory and describes the queuing theoretic model that we use to analyse the potential tension between economic-cost minimization and inspection stringency. To keep the analysis comparative and meaningful, the following section focuses on one of the two economic-cost criteria employed in Batabyal et al. (2005). This cost criterion is the 'average wait of a ship in the port system' or AWS criterion. Next, this section conducts a detailed theoretical and numerical analysis of the aforementioned tension question. The final section concludes and offers suggestions for future research on the subject of this paper.

COST MINIMIZATION AND INSPECTION STRINGENCY

A conceptual framework based on queuing theory

The purpose of queuing theory is to analyse waiting lines or queues mathematically[3]. All queuing models have at least three characteristics. First, there is a probabilistic arrival process. Second, there is a stochastic service process. Finally, there is a fixed number of servers. In the queuing model of our paper, the arrival process is described by a Poisson process. Here, the time between successive arrivals follows an exponential distribution which has the property of being *memoryless*. Hence, it is common to use the letter M to describe the Poisson arrival process.

In general, the service – in our case inspection – times are random and not deterministic. Therefore, it is common to use the exponential distribution—and hence the letter M once again – to model these service times, and this what is done in Batabyal et al. (2005). However, because the primary aim of our paper is to investigate the generality of the central 'tension result' in Batabyal et al. (2005), in what follows, we suppose that the relevant service times are *arbitrarily* distributed. As such, we shall use the letter G to denote the *general* cumulative distribution function of these random service times. Finally, the fixed number of servers – in our case inspectors – is typically denoted by some positive integer, and in the present paper this positive integer is one.

Using the language of queuing theory, the inspection regimes analysed by Batabyal et al. (2005) correspond to the $M/M/1$ and the $M/M/2$ queuing models. In words, Batabyal et al. (2005) have analysed inspection regimes in which the arrival of ships is described by a Poisson process, the time it takes to inspect a ship is exponentially distributed, and the number of inspectors equals either one or two. The inspection regime that we analyse in this paper corresponds to the $M/G/1$ queuing model. This model is more general than the Markovian queuing models in

Batabyal et al. (2005) because the random inspection times are now arbitrarily and not exponentially distributed.

A model of inspections in invasive-species management

Consider a stylized, publicly owned port in a particular coastal part of some nation. Ships with ballast water and/or cargo in containers arrive at this port either to load or to unload cargo, and if they have arrived to load cargo they then transport this cargo to a port in some other part of the world. The arrival of these ships coincides with the arrival of potentially damaging plant and animal species. We assume that the arrival rate of these plant and animal species is proportional to the arrival rate of the ships. Therefore, we shall not model these species directly. Instead, we shall focus on the ships that bring these species to our port by means of either their ballast water or the containers that are used to carry the cargo. The arrival process of the ships in our port represents the arrival process for the queuing-theoretic inspection regimes that we study in this paper. Now, consistent with the discussion in an earlier section, we assume that the ships in question arrive at our port in accordance with a Poisson process with rate λ. Note that all else being equal, a higher λ means two things. First, our port is now handling more cargo or a higher volume of maritime trade. Second, because the arrival rate of the various non-native plant and animal species is proportional to the arrival rate of the ships, a higher λ also means a larger volume of potentially injurious biological organisms and hence a higher likelihood of one or more biological invasions. From this discussion, the reader will note that λ serves as a proxy for both the volume of maritime trade and the likelihood of biological invasions.

Our port manager would like to prevent invasions by the possibly deleterious plant and animal species entering the port under study. Therefore, arriving ships will need to be inspected before they can either load or unload cargo. Ships are inspected on a first-come-first-served basis and an inspector is assigned to each dock in our port, and hence, in what follows, we shall study a *representative* dock inspector's decision problem. In addition, we shall think of the inspection function broadly. For some ships, only the ballast water will need to be inspected. For other ships, only the containers carrying cargo will require inspection. Finally, for a third category of ships, both the ballast water and the containers will need to be inspected. This tells us that inspections will generally require varying amounts of time. To account for this in a general way, we permit the inspection times to be not only random but also to be arbitrarily distributed. The port system under study consists of ships that are being inspected, ships that are waiting to be inspected, the representative dock inspector, and the port manager.

Now, the stringency of inspections is generally an increasing function of the amount of time it takes to complete inspections. Therefore, to model this idea, we assume that there are two possible inspection regimes in our port. In the first or inspection regime (A), the mean inspection time is v_A and the variance of this time is τ_A^2. In the second or inspection regime (B), the average inspection time is v_B

and the variance of this time is τ_B^2. Furthermore, we assume that $v_A > v_B$ and that $\tau_A^2 < \tau_B^2$. These two inequalities tell us that inspection regime A is *more* stringent than inspection regime B. Why? Because relative to regime B, on average, regime A requires that more time is devoted to inspection. In addition, the variability of the time spent inspecting ships in regime A is also less than the variability of the time spent in regime B.

We now have all the necessary parts for our two queuing-theoretic inspection regimes. The reader should note the way in which we have mathematically characterized the central question of this paper: When attempting to prevent a biological invasion by inspecting the ballast water and/or the containers of ships, which inspection regime, A or B, ought our port manager to have in place? We now proceed to the theoretical and the numerical analysis of the inspection regime choice question for the *AWS* cost criteria that we identified in the last paragraph of the introductory section.

THE COST CRITERION

AWS criterion

Inspection activities that result in the prevention of a biological invasion by non-native plant or animal species clearly result in benefits to the citizens of the coastal region under study. However, during the time that arriving ships are being inspected, there is neither loading nor unloading of cargo, and hence in general, economic activity resulting from maritime trade is at a standstill. This temporary stoppage of economic activities imposes costs on the economy of our coastal region. We can measure this cost by computing the average wait of a ship in the port system. In this way of looking at the problem, the longer (shorter) this average wait in the port system or AWS, the larger (smaller) the costs from the interruption of economic activities. Consequently, a port manager who is concerned primarily about the economic costs that are imposed on society by the activities of the representative inspector will want to keep AWS as low as possible. In contrast, a port manager who worries more about the potential damage to society from a biological invasion will want to have the more stringent or inspection regime (A) in place. In what follows, we assume that our port manager has this AWS (proxy for economic cost) criterion in mind when (s)he is choosing between regimes A (more stringent) and B (less stringent).

Let us now calculate AWS for the two $M/G/1$ inspection regimes that we are analysing in this paper. From equation 3.17 in Taylor and Karlin (1998, p. 563) we conclude that the two expressions we seek are given by

$$AWS_A = v_A + \{\lambda(\tau_A^2 + v_A^2)\}/\{2(1 - \lambda v_A)\} \quad \text{and} \quad (1)$$

$$AWS_B = v_B + \{\lambda(\tau_B^2 + v_B^2)\}/\{2(1 - \lambda v_B)\}$$

respectively. We know that inspection regime A is more stringent than inspection regime B. Mathematically, this means that $v_A > v_B$ and $\tau_A^2 < \tau_B^2$. Using the first inequality we conclude that $2(1 - \lambda v_A) < 2(1 - \lambda v_B)$. However, because $(\tau_A^2 + v_A^2)$ may be bigger or smaller than $(\tau_B^2 + v_B^2)$, knowing that $v_A > v_B$ and that $\tau_A^2 < \tau_B^2$ does *not* allow us to conclude anything definitively about the relative magnitudes of AWS_A and AWS_B. Put differently, when our port manager authorizes the use of the more stringent A inspection regime in the port under study, it is *not* necessarily the case that economic costs measured by the AWS criterion will be higher. This tells us that when one works with the $M/G/1$ queuing model, in the general case, there may or may not be a tension between economic-cost minimization and inspection stringency. Hence, a key finding in Batabyal et al. (2005) does *not* generalize to the case in which the inspection times are arbitrarily and not exponentially distributed.

Now, to show that there is no straightforward resolution of the tension question, we conduct an exercise with specific numerical values for the various model parameters of interest. To this end, let the arrival rate of ships be $\lambda = 1$ per unit time. Also assume that the parameters of the two inspection regimes are $(v_A, \tau_A^2) = (0.5, 0.2)$ and $(v_B, \tau_B^2) = (0.4, 0.9)$. Then, using equation (1), it is easy to see that $AWS_A = 0.5 + \{0.45\lambda/(2 - \lambda)\}$ and $AWS_B = 0.4 + \{1.06\lambda/(2 - 0.8\lambda)\}$. When $\lambda = 1$, these expressions reduce to $AWS_A = 0.95$ and $AWS_B = 1.28$.

These two expressions lead to two conclusions. First, examining the expression for $AWS_A(AWS_B)$ we see that as the arrival rate of ships λ approaches 2 (2.5), economic costs measured by the $AWS_A(AWS_B)$ criterion become infinitely large. In other words, there is a definite upper limit on the volume of maritime trade that our port can handle, and when this limit is approached, the economic costs of inspections become prohibitively large. Second, when the parameters of our model take on the values specified in the previous paragraph, our port manager will prefer the more stringent inspection regime (A) over the less stringent inspection regime (B). This is because when this more stringent regime is in place, the economic costs of inspections are lower, i.e., $AWS_A = 0.95 < 1.28 = AWS_B$. We have just identified a case in which there is *no* tension between economic-cost reduction and the stringency of inspections or biological-invasion damage control. It is instructive to analyse this tension question in three different ways and we now proceed to this tripartite analysis sequentially.

The tension question in terms of the volume of maritime trade

Let us study – for the model parameter values specified above – the dependence of the magnitude of the economic costs (measured by AWS) on our proxy for the volume of maritime trade, i.e., on λ. We begin by equating the two expressions for AWS obtained previously. Doing this and then simplifying the resulting expression gives us a quadratic equation in λ. That equation is

$$0.78 \, \lambda^2 - 1.58\lambda + 0.4 = 0. \tag{2}$$

The two solutions to equation (2) are $\lambda_1^* = 1.73$ and $\lambda_2^* = 0.30$. Figure 1 plots the economic-cost criterion AWS on the vertical axis against selected values of the arrival rate of the ships or λ on the horizontal axis. Looking at Figure 1, the reader can easily verify that when $\lambda < 0.30$ or when $\lambda > 1.73$, our port manager will prefer to have inspection regime B in place rather than inspection regime A. Why? Because inspection regime B results in lower economic costs as measured by the AWS criterion. Put differently, when $\lambda < 0.30$ or when $\lambda > 1.7$, there *is* a tension between economic-cost reduction and the stringency of inspections or biological-invasion damage control. In contrast, for all λ in the closed interval *[0.30,1.73]*, there is *no* tension between economic-cost reduction and inspection stringency.

The tension question in terms of the average inspection times

Very stringent inspections are time-consuming and they tend to increase the magnitude of the AWS criterion. Therefore, intuitively speaking, we would expect the answer to the question about whether there is or isn't a tension between economic-cost reduction and inspection stringency to be clearly related to the means and the variances of the A and the B inspection regimes. Therefore, in this section, we numerically investigate the functional dependence of AWS on the means (v_A, v_B) of the two inspection regimes, and in the next section, we shall conduct a similar exercise from the standpoint of the two variances (τ_A^2, τ_B^2).

We know that $v_A > v_B$ and that $\tau_A^2 < \tau_B^2$. Further, in our subsequent numerical analysis, we assume that $v_A = a v_B, a > 1$, and that $\tau_A^2 = b \tau_B^2, 0 < b < 1$. In words, the means and the variances of the two inspection regimes are linearly related to each other and the two constants of proportionality *(a,b)* satisfy certain

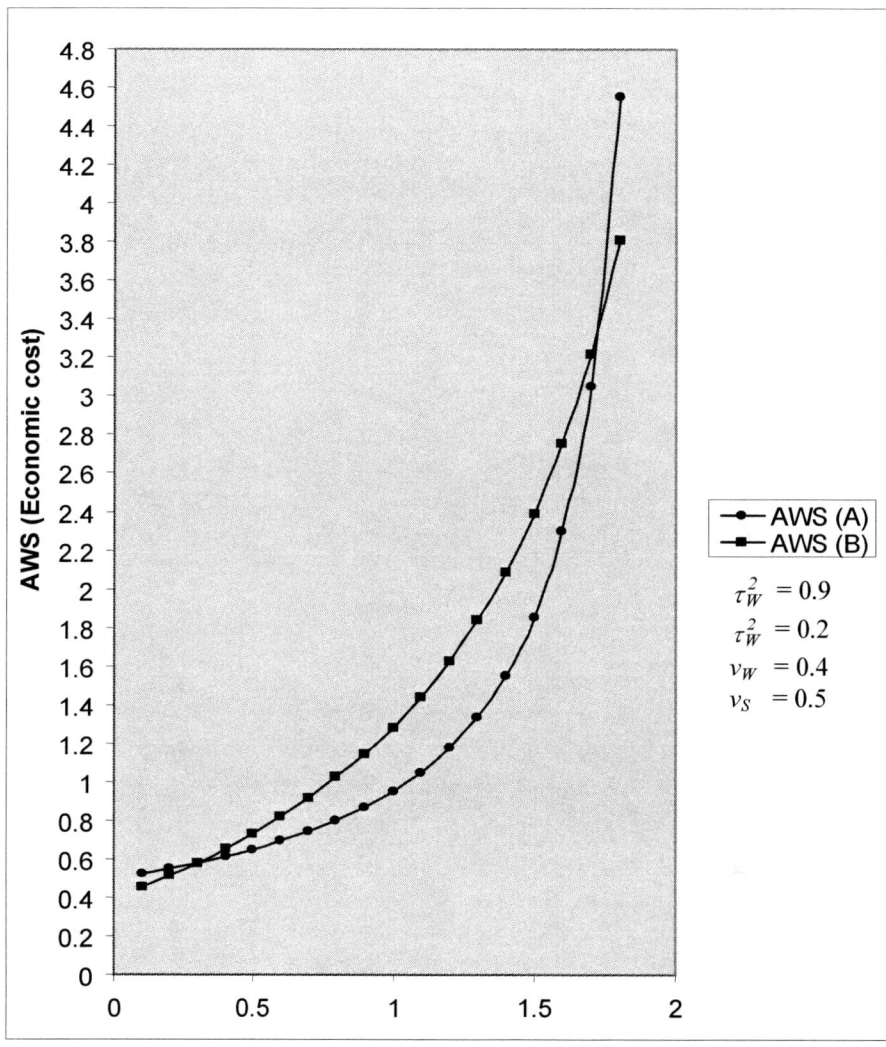

Figure 1. *AWS for inspection regimes A and B as a function of lambda (arrival rate of ships)*

straightforward restrictions. Specifically, because $v_A > v_B$ we have $a > 1$. Similarly, because $\tau_A^2 < \tau_B^2$ it makes sense for b to lie in the open interval $(0,1)$. The reader should think of the parameter a as a measure of the difference in the stringencies of the two inspection regimes A and B. That is, as a increases, inspection regime A becomes more stringent than the B inspection regime. Similarly, the parameter b describes the difference in the variability of the two inspection regimes. Hence, as b approaches zero inspection regime A becomes more reliable relative to inspection regime B and as b approaches unity regime A becomes less reliable relative to regime B.

Now, using the parameter values from the previous section, we have $\lambda = 1, v_B = 0.4, \tau_B^2 = 0.9$, and setting b at its mid-point, i.e., $b = 0.5$, we obtain $v_A = 0.4a$ and $\tau_A^2 = 0.5\tau_B^2$. Using these values of the various parameters in equation (1), we get $AWS_B = 1.2833$, and AWS_A is a function of the parameter a and is given by $AWS_A = 0.4a + (0.45 + 0.16\,a^2)/(2 - 0.8a)$. Setting these two values equal gives us a quadratic equation in a, and that equation is

$$0.16\,a^2 - 1.83a + 2.12 = 0. \tag{3}$$

The two solutions to equation (3) are $a_1^* = 1.31$ and $a_2^* = 10.11$. Now for $AWS_A = 0.4a + (0.45 + 0.16\,a^2)/(2 - 0.8a)$ to be positive we must have $a < 2.5$. This tells us that $a_2^* = 10.11$ is an inadmissible solution in our case and we are left with $a_1^* = 1.31$ as the only economically meaningful solution to equation (3).

Figure 2 plots the economic-cost criterion AWS on the vertical axis against selected values of a on the horizontal axis. Looking at figure 2 we see that when $a = 1.31$ our port manager is indifferent between the two inspection regimes. Further, for all $a < 1.31$, the use of the more stringent A inspection regime results in lower economic costs as measured by the AWS criterion. Finally, for all $a > 1.31$, the use of the less stringent B inspection regime leads to lower economic costs. This tells us that when $a > 1.31$ there *is* a tension between economic-cost reduction and biological-invasion damage control. In contrast, when a lies in the interval $(1, 1.31]$ there is *no* tension between economic-cost reduction and inspection stringency.

The tension question in terms of the variances of the inspection times

Many stochastic models exhibit significant qualitative differences depending on the variability of the underlying distributions. Therefore, we now numerically study the functional dependence of AWS on the variances – τ_A^2, τ_B^2 – of the A and the B inspection regimes. Recall that we have $\tau_A^2 < \tau_B^2$ and $v_A > v_B$. Also, we once again have $v_A = a v_B, a > 1$, and $\tau_A^2 = b \tau_B^2, 0 < b < 1$. The interpretation of a and b is as indicated in the previous section. Using the previous section's

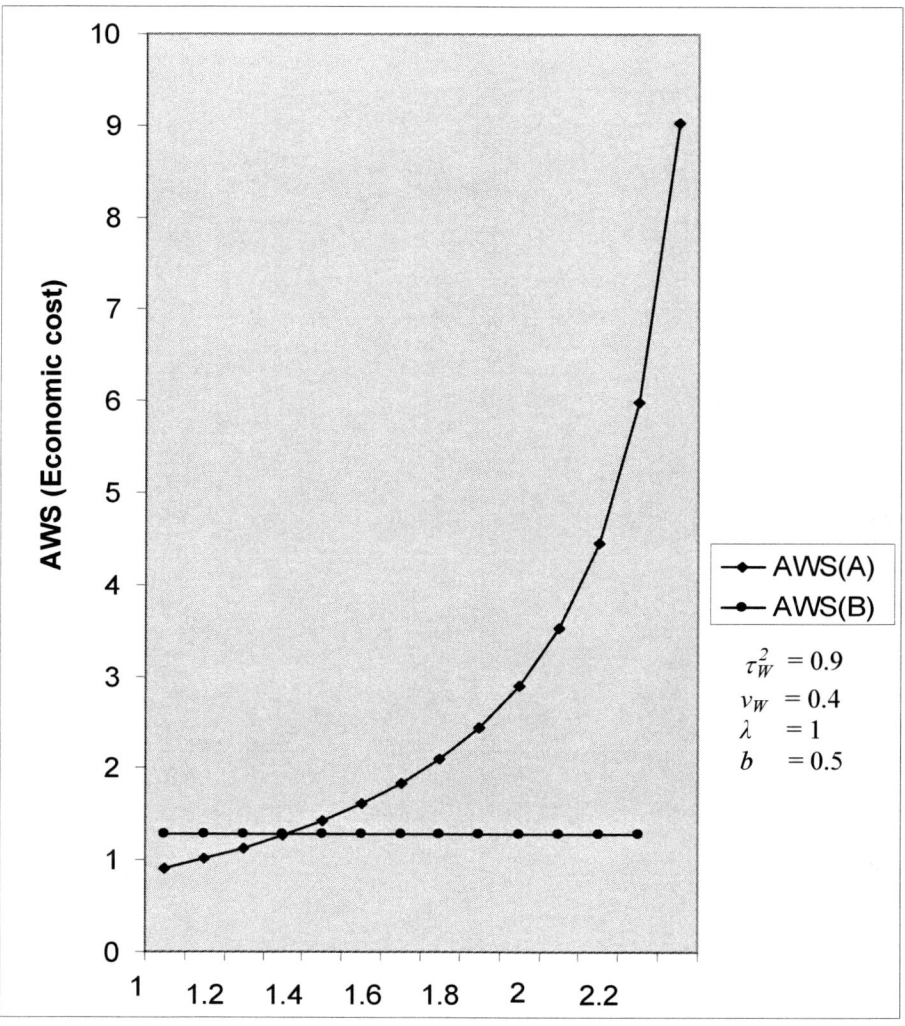

Figure 2. *AWS for inspection regimes A and B as a function of a (the relative stringency of inspections)*

parameter values, we have $\lambda = 1, v_B = 0.4, \tau_B^2 = 0.9$. Now setting $a = 2$ and using equation (1), we get $AWS_B = 1.2833$, and we can write AWS_A as a linear function of b; that function is $AWS_A = 2.4 + 2.25b$. Looking at these two values of the economic-cost criterion it is obvious that there is no value of b for which our port manager would be indifferent between the two inspection regimes being considered.

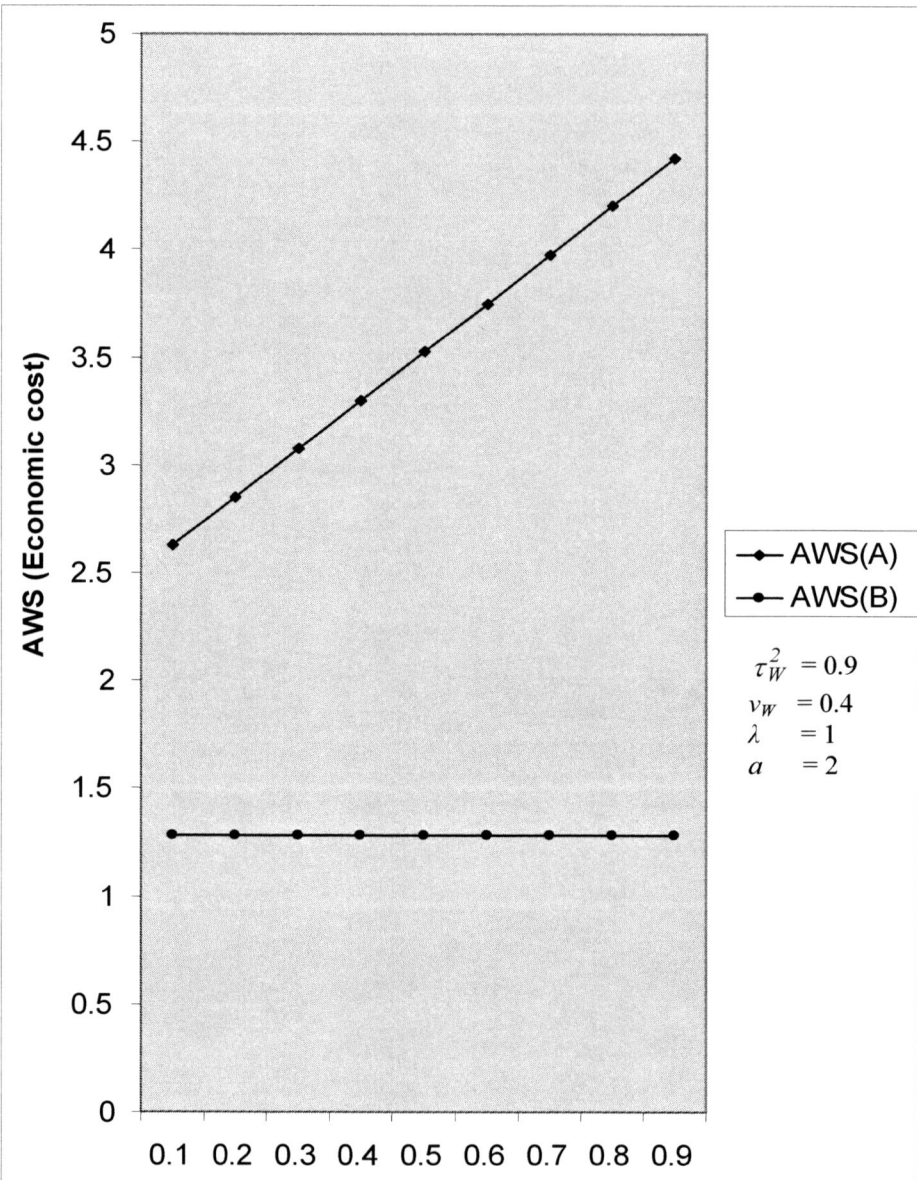

Figure 3. *AWS for inspection regimes A and B as a function of b (the relative variability of inspections)*

Figure 3 plots the economic-cost criterion AWS on the vertical axis against alternate values of b on the horizontal axis. Looking at Figure 3 we see that AWS

is *always* lower when the less stringent B inspection regime is used to inspect arriving ships in our port. Put differently, for *all* values of b – which measures the difference in the variability of the two inspection regimes – there *is* a tension between economic-cost minimization and biological-invasion damage control.

Our analysis thus far in this section leads to three conclusions. First, our theoretical examination shows that the question as to whether there is or isn't a tension between economic-cost reduction and inspection stringency cannot be resolved unambiguously. Second, for many possible values of a and for *all* possible values of b there *is* a tension between economic-cost reduction and inspection stringency. Finally and in contrast with the second point, for several possible values of λ or the 'volume of maritime trade' parameter, there is *no* tension between economic-cost reduction and inspection stringency.

CONCLUSIONS

Maritime trade in goods by means of ships often results in biological invasions of novel habitats by non-native plant and animal species. Therefore, if an apposite authority such as a port manager's aim is to preclude biological invasions, then (s)he must inspect arriving ships for possibly injurious biological organisms. Given this context, we used the $M/G/1$ queuing model to investigate the generality of a central result in Batabyal et al. (2005). This result tells us that there is a tension between economic-cost reduction and inspection stringency or biological-invasion damage control. Our theoretical analysis showed that in the more general case in which inspection times are arbitrarily and not exponentially distributed, there is *no* definite answer to the question as to whether there is or isn't a tension between economic-cost minimization and biological-invasion damage control. In addition, our numerical analysis with arbitrary values for the key model parameters identified specific ranges for these parameters for which there is a tension between economic-cost minimization and biological-invasion damage control. The upshot of our combined theoretical and numerical analysis is this: whether or not there is tension depends very much on the specifics of a particular situation.

For real applications, we would need to obtain values for the arrival rate of ships and the means and the variances of specific inspection regimes. In the USA, information about the arrival rate of ships can be obtained from the administrative offices of individual ports such as Long Beach and, on occasion, from governmental agencies such as the Office of Mobile Sources of the Environmental Protection Agency (EPA). Similarly, information about actual inspections in the USA can be obtained from documents that are periodically produced by the Congressional Research Service and from the Animal and Plant Health Inspection Service (APHIS).

The analysis in this paper can be extended in a number of directions. We now suggest two possible extensions of this paper's research. First, it would be useful to use the $M/G/1$ queuing model to set up and solve an optimization problem in which the port manager chooses the mean inspection time to optimize a specific objective function. Second, it would also be useful to analyse the tension question of this paper in a scenario in which excessively long inspection times result in some ships not entering the port under study. This would involve the analysis of a $M/G/1$ queuing model with 'balking'. Studies of maritime-trade-driven biological invasions that incorporate these aspects of the inspection function into the analysis will provide additional insights into a management problem that has considerable economic and biological implications.

NOTES

[1] Batabyal acknowledges financial support from the USDA's PREISM program by means of Cooperative Agreement 43-3AEM-4-80100 and from the Gosnell endowment at RIT. The usual disclaimer applies.
[2] For more on this, see Nunes and Van den Bergh (2004), Yang and Perakis (2004) and Batabyal and Beladi (2006).
[3] Textbook accounts of queuing theory can be found in Taylor and Karlin (1998), Ross (2003) and Tijms (2003).

REFERENCES

Barbier, E.B., 2001. A note on the economics of biological invasions. *Ecological Economics,* 39 (2), 197-202.
Batabyal, A.A., 2004. A research agenda for the study of the regulation of invasive species introduced unintentionally via maritime trade. *Journal of Economic Research,* 9 (2), 191-216.
Batabyal, A.A. and Beladi, H., 2006. International trade and biological invasions: a queuing theoretic analysis of prevention problem. *European Journal of Operational Research,* 170 (3), 758-770.
Batabyal, A.A., Beladi, H. and Koo, W.W., 2005. Maritime trade, biological invasions, and the properties of alternate inspection regimes. *Stochastic Environmental Research and Risk Assessment,* 19 (3), 184-190.
Batabyal, A.A. and Nijkamp, P., 2005. On container versus time based inspection policies in invasive species management. *Stochastic Environmental Research and Risk Assessment,* 19 (5), 340-347.
Cox, G.W., 1993. *Conservation ecology.* William C. Brown Publishers, Dubuque.
Eiswerth, M.E. and Johnson, W.S., 2002. Managing nonindigenous invasive species: insights from dynamic analysis. *Environmental and Resource Economics,* 23 (3), 319-342.
Eiswerth, M.E. and Van Kooten, G.C., 2002. Uncertainty, economics, and the spread of an invasive plant species. *American Journal of Agricultural Economics,* 84 (5), 1317-1322.
Horan, R.D., Perrings, C., Lupi, F., et al., 2002. Biological pollution prevention strategies under ignorance: the case of invasive species. *American Journal of Agricultural Economics,* 84 (5), 1303-1310.
Nunes, P.A.L.D. and Van den Bergh, J.C.J.M., 2004. Can people value protection against invasive marine species? Evidence from a joint TC-CV survey in the Netherlands. *Environmental and Resource Economics,* 28 (4), 517-532.
Olson, L.J. and Roy, S., 2002. The economics of controlling a stochastic biological invasion. *American Journal of Agricultural Economics,* 84 (5), 1311-1316.
OTA, 1993. *Harmful non-indigenous species in the United States.* Office of Technology Assessment, United States Congress, Washington.
Parker, I.M., Simberloff, D., Lonsdale, W.M., et al., 1999. Impact: toward a framework for understanding the ecological effects of invaders. *Biological Invasions,* 1 (1), 3-19.
Perrings, C., Williamson, M. and Dalmazzone, S. (eds.), 2000a. *The economics of biological invasions.* Elgar, Cheltenham.
Perrings, C., Williamson, M. and Dalmazzone, S., 2000b. Introduction. *In:* Perrings, C., Williamson, M. and Dalmazzone, S. eds. *The economics of biological invasions.* Elgar, Cheltenham.
Pimentel, D., Lach, L., Zuniga, R., et al., 2000. Environmental and economic costs of nonindigenous species in the United States. *BioScience,* 50 (1), 53-65.

Ross, S.M., 2003. *Introduction to probability models.* 8th edn. Academic Press, San Diego.

Taylor, H.M. and Karlin, S., 1998. *An introduction to stochastic modeling.* 3rd edn. Academic Press, San Diego.

Tijms, H.C., 2003. *A first course in stochastic models.* John Wiley and Sons, Chichester.

USDA-APHIS, 2000. *Pest risk assessment for importation of solid wood packing materials into the United States.* Animal and Plant Health Inspection Service.

Vitousek, P.M., D'Antonio, C.M., Loope, L.L., et al., 1996. Biological invasions as global environmental change. *American Scientist,* 84 (5), 468-478.

Yang, Z. and Perakis, A.N., 2004. Multiattribute decision analysis of mandatory ballast water treatment measures in the US Great Lakes. *Transportation Research. Part D. Transport and Environment,* 9(1), 81-86.

CHAPTER 4

DESIGNING OPTIMAL PHYTOSANITARY INSPECTION POLICY

A conceptual framework and an application

ILYA V. SURKOV[#], ALFONS G.J.M. OUDE LANSINK[#], WOPKE VAN DER WERF[##] AND OLAF VAN KOOTEN[###]

[#] *Business Economics Group, Wageningen University, Postbus 8130, 6700 EW Wageningen, The Netherlands. E-mail: Ilya.Surkov@wur.nl*
[##] *Crop and Weed Ecology Group, Wageningen University, Postbus 430, 6700 AK Wageningen, The Netherlands*
[###] *Horticultural Production Chains Group, Wageningen University, Marijkeweg 22, 6709 PG Wageningen, The Netherlands*

Abstract. Optimal allocation of available resources to minimize quarantine risks related to international trade is a problem facing plant protection agencies worldwide. In this paper a model of budget allocation to minimize quarantine risks is developed. Theoretical conditions that budget allocation should satisfy are derived. These conditions imply that optimal allocation of resources is achieved when the marginal pest risks are equalized across risky pathways. Furthermore, an empirical model of budget distribution is developed. In the empirical model, the protecting agency wants to minimize the expected number of infested ornamental plants imported in a given country. The model is parameterized using data on import of ornamental commodities, the associated quarantine risks and costs of import phytosanitary inspections pertaining to the Netherlands.

The results of the empirical model suggest that under specific assumptions (such as constant risk) greater risk reduction can be achieved by allocating larger funds to inspection of riskier pathways, and less or no funds to less risky pathways. The protecting agency has to trade off the risks from pathways that vary in terms of risk.

Keywords: optimal inspection; quarantine pest; ornamental plants

INTRODUCTION

Phytosanitary import inspection is an important component of quarantine policies worldwide. In many instances, import inspection is the only real and last barrier where exotic quarantine plant pests brought in together with imported commodities can be intercepted. The inspection capabilities of the responsible agencies are, however, under a constant pressure of the ever-growing volumes of importing commodities. There is evidence that in some countries the resources of the

A.G.J.M. Oude Lansink (ed.), New Approaches to the Economics of Plant Health, 39–54.

quarantine agencies are already lagging behind the increasing volumes of import (National Research Council 2002; Everett 2000). In addition, the broad assortment and origins of incoming consignments diversify phytosanitary risks and complicate inspection tasks of responsible agencies.

The economic rationale calls for the best use of available inspection resources, including monetary and human resources. More attention should therefore be paid to development of inspection policies in which scarce resources are allocated optimally and risks associated with import of various commodities are minimized. The treatment of this issue in the economic literature so far has been limited. Relevant studies focus on economics of controlling and preventing biological invasions (e.g. Horan et al. 2002; Saphores and Shogren 2005; Barbier 2001; Olson and Roy 2002), which is a somewhat broader phenomenon. A most relevant study on the economics of import inspection is a recent paper (Batabyal and Beladi in press) in which queuing theory is applied to analyse optimal allocation of resources for inspection of cargo ships. The general feature of these studies is that, though they provide theoretical conditions for optimal resource allocation, numerical examples are lacking. As a result it remains unclear how these theoretical conditions may be translated into practical decision making.

This paper adds an applied focus to the problem of optimal allocation of quarantine resources. Specifically, the main question addressed in the current work is: how can available resources be allocated to inspection of imported commodities such that the phytosanitary risks associated with these imports are minimized? To answer this question, first, a theoretical model of optimal budget allocation is proposed. In this model, the decision maker – the Quarantine Agency of an importing country – faces a problem of resource allocation to minimize quarantine risks stemming from different pathways (defined as commodity–country combinations). Based on this theoretical model, the empirical model is then developed. In this model the Agency wants to minimize the number of infested plants imported into the country. Data from the phytosanitary import inspections of ornamentals imported into the Netherlands were used to parameterize the model. The results of the optimal budget allocation are then presented. The paper concludes with a discussion.

THEORETICAL MODEL

Consider an importing country H that imports j commodities from i exporting countries in period t. Each of the j commodities may host k quarantine pests, currently not present in H. The Quarantine Agency considers the presence of any of these pests inside H as equally (economically) unacceptable. The Agency thus has no specific aversion towards a specific pest and treats all pests equally. The latter assumption has a simplifying implication that the Agency applies the same quarantine measures to all ij pathways. The only phytosanitary measure applied by the Agency is the visual inspection of incoming consignments along each of the ij pathways. For inspection, a sample of a pre-defined size is taken from every consignment. If at least one specimen of a quarantine organism is found in a sample, the entire consignment is rejected for import. Otherwise, the consignment is freely imported.

Denote the quarantine risk associated with the ijth pathway in period t as $r_{ij}^t \geq 0$. (The superscript implies that risk is period-specific; however, as the discussion henceforth is confined to a single period t, the superscript will be omitted.) Assume that r is measured in units that the Agency deems appropriate to reflect the quarantine risk associated with imported commodities. In reality, r may be expressed, e.g., as the expected economic costs due to pest incursion, the probability of pest establishment in H, the number of infested plant units or any other 'real' risk metric. The total import quarantine risk in period t is given by $R_t = \Sigma_{ij} r_{ij}$, assuming that risks from different pathways are not correlated.

Developing its risk management policy (i.e. import inspection), the Agency realizes that no inspection measures can reduce risk to zero. Hence, the Agency may impose a risk threshold below which risk is considered acceptable; consequently, commodities satisfying this threshold are imported without inspection. The Agency may choose to set the *total* risk threshold $\overline{R}$ or *individual* pathway risk threshold $\overline{r}$ [1] In the former case, total risk from all commodities should be lower than or equal to $\overline{R}$, i.e. $R_t \leq \overline{R}$; likewise, in the latter case, pathways' risks should not exceed $\overline{r}$, i.e. $r_{ij} \leq \overline{r}$. It is, however, more difficult to maintain $R_t \leq \overline{R}$ than $r_{ij} \leq \overline{r}$ constraint because management efforts should change with fluctuation in the trade volumes (Bigsby 2001). With the individual pathway constraint, management effort is constant. Henceforth, we assume that the Agency imposes an individual pathway risk constraint $\overline{r}$. The inspection measures applied by the Agency are consistent with the imposed constraint; i.e., the sampling procedure is such that the acceptable level of risk is maintained.

Inspection and sampling are, of course, costly. To reflect this, an inspection budget $b_{ij} \geq 0$ is allocated to each pathway. As a result, the quarantine risk per pathway is a function of the allocated budget, i.e. $r_{ij} = r(b_{ij})$. Assume that $r'(b) < 0$ and $r''(b) > 0$, so that risk is decreasing with budget, but the marginal risk-reducing effect of an extra unit of budget is decreasing. In relation to visual inspection, this implies that an extra inspection effort reduces the quarantine risk; however, subsequent inspection efforts decrease risk less than proportionally, reflecting the increasing difficulties in pest detection.

We can now formulate the optimization problem of the Agency. The relevant objective is to minimize the inspection costs subject to the acceptable risk constraint. The minimization problem (model *MB*) therefore reads as:

$$\text{Minimize } B = \sum_{ij} b_{ij} \tag{1}$$

subject to $r(b_{ij}) \leq \overline{r} \qquad \forall\ i,j,$

$b_{ij} \geq 0.$

Because the risk constraint may not be binding, the solution to (1) will be given by Kuhn-Tucker conditions (Chiang 1984). The first-order conditions (FOC) to this problem are given by $\dfrac{1}{\varphi_{ij}} = r'(b_{ij}) < 0$ and $r'(b_{ij}) = \bar{r}$, where φ is the Lagrange multiplier associated with the ijth constraint. The FOCs imply that the optimal budget allocation is the one that makes individual pathway risks exactly equal to the constraint; at the same time, for pathways with initial risks strictly below $\bar{r}$, the budget should optimally be zero. Note that the Agency with unlimited budget may alternatively insure itself from all risks above $\bar{r}$ by trivially applying the same inspection procedures for all pathways, irrespective of actual r_{ij}'s. The spending of resources in this case will be clearly suboptimal as pathways with risks strictly lower than $\bar{r}$ will be inspected.

More relevant for import quarantine decision making is the situation when the budget is limited. Note that although the budget itself may be sufficient (because in most cases importers pay inspection fees), the complete inspection of all pathways may be unfeasible, e.g., due to the lack of qualified employees or the lack of inspection premises. Thus, with limited budget B (in period t), the Agency solves the following program (model MR):

$$\text{Minimize } R_t = \sum_{ij} r\ (b_{ij}) \tag{2}$$

subject to $\sum_{ij} b_{ij} \le B \qquad \forall\ i,j,$

$b_{ij} \ge 0.$

The constraint in fact should be binding in the optimum because it is always preferable to spend the budget 'a little bit more' to reduce risk marginally. Hence, the FOC is given by $r'(b_{ij}) = \lambda$ implying that in the optimum budget should be allocated such as to equalize the marginal pest risks across all pathways. The Lagrange multiplier λ is the 'shadow price' (Chiang 1984) of the budget constraint; it shows how the total risk will decrease (because $r'(b_{ij}) < 0$) when the budget constraint is relaxed. The limited budget in this model implies that in the optimal solution not all pathways may be inspected at the level satisfying $\bar{r}$. As a result, quarantine risks from some pathways may exceed the acceptable level $\bar{r}$.

Altogether, the results of MB and MR models provide an indication of how the Agency should allocate its resources optimally. As was mentioned in the Introduction, most quarantine agencies worldwide face binding budget constraints. Hence, the empirical model presented in the next section is based on the MR model.

EMPIRICAL MODEL

To translate a conceptual MR model into an empirical one, firstly, we need to specify a concrete objective function to be minimized – i.e. assume a specific risk function r, and secondly, establish relations between the costs of inspections (i.e. b_{ij}) and their efficacy (i.e. $r'(b)$). Obviously, for the model to yield practical insights, assumed empirical specifications should resemble the actual import inspection practice.

Given our earlier assumption that the Agency has no bias against specific pests, the relevant objective function is to minimize the expected number of infested commodity units imported into H. For concreteness, assume that the imported commodity is the ornamental materials for propagation (for example, cuttings or small plants for propagation; hereafter, simply 'plant') of j ornamental species. We thus implicitly assume that each infested plant may lead to realization of a quarantine risk in H with constant and independent (of other infested plants) probability of success. Given the limited budget B, the objective of the Agency is to:

$$\text{Minimize } E(N) = \sum_{ij} N_{ij}(b_{ij}) \qquad (3)$$

subject to $\sum_{ij} b_{ij} \le B$

$b_{ij} \ge 0,$

where $N_{ij}(b_{ij})$ is the expected number of infested plants imported along the ijth pathway after import inspection. Specifically, it is given by:

$$N_{ij}(b_{ij}) = V_{ij} p_{ij} \alpha(b_{ij}), \qquad (4)$$

where V_{ij} is the volume of plants imported along ijth pathway in period t, p_{ij} is the proportion of infestation with quarantine pests in the total population of ornamental plant j in country i and $\alpha(b_{ij})$ is the probability that inspection will fail to detect at least one infested plant in the infested consignment. The probability of inspection failure is assumed to be decreasing and convex in the inspection budget, i.e. $\alpha'(b)<0$ and $\alpha''(b)>0$. V_{ij} is defined as $\sum_{z=1}^{Z} h_z^{ij}$ where h_z^{ij} is the size of the zth consignment. The proportion of infestation p_{ij} is estimated according to the following formula:

$$p_{ij} = \frac{u_{ij}}{v_{ij}} p_{\text{inf}}, \qquad (5)$$

where v_{ij} is the total volume of commodity imported along the ijth pathway in periods preceding t, u_{ij} is the total volume of consignments found infested with quarantine pests during import inspection for the same periods, and p_{inf} represents the assumed percentage share of u_{ij} actually infested with quarantine pests (see section 'Data' for explanation).

The Agency may vary the intensity of visual inspection by taking larger samples, hence lowering the probability $\alpha(b_{ij})$ that an infested plant remains undetected. We assume that detection probability is independent of the pest type and the type of propagation material. Statistically, the probability of detecting an infested plant in a given consignment is a function of the proportion of infestation p_{ij} and the sample size s (when s is small relative to consignment size), assuming binomial distribution of infested plants. Because the proportion of infestation is always unknown, the common convention is to assume a certain critical level of infestation p_c below which a consignment is deemed free from quarantine organisms (e.g. Kuno 1991; Couey and Chew 1986). The resulting sample size is a function of this threshold and the acceptable level of error α. The exact formula is given by Kuno (1991):

$$s = \frac{\ln(\alpha)}{ln(1 - p_c)} . \tag{6}$$

Equation (6) implies that s is decreasing in α, that is, a higher error probability is associated with smaller sample; also, s is decreasing in p_c reflecting that a smaller sample is required when the Agency is prepared to tolerate higher infestation level in a consignment. Equation (6) suggests that the pathway risk accepted by the Agency (i.e. $\bar{r}$) is a function of both α and p_c. For the purposes of the current model we assume that the Agency fixes p_c and may vary sample size to achieve lower error probability. Specifically, we assume $p_c=0.005$. This is a common maximum infection level required by quarantine agencies worldwide, e.g., in New Zealand (Ministry of Agriculture and Forestry 2006) and in the countries that are members of the European Plant Protection Organisation (EPPO 2005). With p_c fixed, equation (6) can be solved for different α's.

Next, we relate the costs of inspection and sample size. Obviously, larger samples require more inspection time and are therefore more costly. We assume that inspection time is measured in 15-minute intervals during which the inspector may examine a fixed number of plants (equal to the sample size). Within 30 minutes, the inspector may inspect a larger sample, and so on. His productivity is however diminishing. Data about the costs of inspection came from the Dutch Plant Protection Service (PD) that charges a fixed rate for every 15 minutes of inspection. The costs for 0-105-minute inspections, together with corresponding error levels and sample sizes, are shown in Table 1 below.

Table 1. *Relation between sample size, error level α, inspection length and sample costs* ($p_c=0.005$)

Inspection length, minutes	Sample size, units	α	Inspection costs ('15 minutes' fee + 'call out' fee)*, euros
0	0	1,0000	0
15	300	0,2223	61.61
30	570	0,0574	83.28
45	825	0,0160	104.95
60	1065	0,0048	126.62
75	1260	0,0018	148.29
90	1434	0,0008	169.96
105	1587	0,0004	191.63

*callout fee: 39.94 euros, '15 minutes' fee: 21.67 euros.
Source: (Plantenziektenkundige Dienst 2005)

The chosen inspection lengths were based on presumption of the 'reasonable' length. One might argue that the inspection lengths longer than 60 minutes are unfeasible in practice; nevertheless, for completeness, longer inspection intervals were included. The second column shows the assumed sample sizes that can be inspected within a corresponding inspection time. Note that the sample size is a concave function of the inspection time. This reflects the assumed diminishing marginal productivity of an inspector. The α's are obtained by solving (6) for fixed p_c and s. Examining the relation between the last two columns one finds that α is decreasing and convex in inspection costs (consistent with our earlier assumptions about $\alpha(b_{ij})$).

DATA

In the empirical model, nine pathways are considered: three countries each exporting three ornamental species (propagating materials) to the Netherlands. Countries are indexed as A, B and C for confidentiality reasons. The exact pathways are the following: country A, Chrysanthemum, Rose and Dianthus; country B, Chrysanthemum, Dianthus and Impatiens; and country C, Chrysanthemum, Yucca and Dracaena. (Henceforth, unique pathways will be referred to by the name of the underlying ornamental species only (i.e. Rose, Yucca, Impatiens and Dracaena); for the remaining pathways a letter denoting the country index will be added to the species name, e.g., DiathusA.) The chosen pathways give a representative sample of the important channels of ornamental materials for propagation imported into the Netherlands. So, for example, in 1998-2001, the six ornamental species chosen for the model accounted for more than 81 % of Dutch import of ornamental plants and propagating materials. (The total number of imported ornamental species for the same period was approximately equal to 1,200.) Chrysanthemum and Dianthus contributed with by far the largest shares: 66.8 % and 11.6 %, respectively. Remaining pathways' shares vary between 0.3 % and 2.7 %. The exporting countries were selected as important suppliers of respective ornamental species. For example, country A accounted for 30 % of Chrysanthemum exports and 38 % of Rose exports; country B supplied 18 % of Dianthus and 43 % of Impatiens; finally, country C exported 11 % of Chrysanthemum and dominated the export of Dracaena with an 84 % share. At the same time, for non-unique pathways (e.g., Chrysanthemum), there is a significant variation in imported volumes between exporting countries (see next paragraph). This circumstance plus the differences in historical findings of quarantine organisms (see below) were the final criteria based on which the pathways were chosen. Data on import volumes and results of import phytosanitary inspections were obtained from the database of inspection reports composed by the PD inspectors in the period 1998-2001. It should be noted that information in the database was presented at the *lot* level, with a lot typically representing a collection of imported plants or plant materials of a given species coming from a given country. A *consignment*, on the other hand, may consist of different lots covered by a single phytosanitary certificate (FAO 2006). For the purposes of the data analyses we consider each lot in the database as a single consignment.

Table 2 presents both historical data on import volumes and findings of quarantine organisms[2] and input data for the model. Consider first historical import data. Consignment-wise, Dianthus and Dracaena were imported in largest numbers compared to other ornamental species. In terms of the average consignment size, Chrysanthemum is leading. Yet for both parameters, there is substantial intra-pathway variation. For the model, the average volume of import expected in a given period t along the ijth pathway, V_{ij}, can be obtained by a straightforward multiplication of the number of consignments and their average size. It is, however, unlikely that all consignments will have the same size. We chose a pragmatic approach to represent this variation in size splitting the historical distribution of consignment sizes into discrete intervals, represented by the lower 5 %, 5-25 %, 25-50 %, 50-75 %, 75-95 % and upper 95 % percentiles. The expected number of consignments of a specific size was thus split according to these percentiles. This transformation is not shown due to space limitations but can be obtained upon request. The important issue to keep in mind is that the increasing percentile implies a greater consignment size (i.e. lower 5 % percentile gives 5 % of the smallest consignments, 5-25 % percentile represents 20 % of consignments of larger size, etc.). For further reference, the total number of plants to be imported (calculated for average consignment sizes) is approximately 671 million.

Data on findings of quarantine pests reveal that consignments of Dianthus have the largest relative and absolute rejection rate. (It is assumed that: 1) inspection procedures applied were the same for all pathways, and 2) all infested consignments were detected.) Most notably, DianthusA has the highest rejection rate among all pathways, suggesting that the underlying pathway is the most risky from the quarantine perspective. The second highest rejection rate among ornamental species pertains to consignments of Chrysanthemum. Finally, consignments of Dracaena have the lowest positive rejection rate. The remaining pathways (i.e. Rose, Impatiens and Yucca) had a zero rejection rate suggesting that these are the safest pathways from a phytosanitary perspective.

Table 2. *Data and parameter values for the model**

Parameter	Ornamental species					
	Chrysanthemum	Rose	Dianthus	Yucca	Dracaena	Impatiens
A						
Consignments imported during 1998-2001	2,375	153	2,909			
Average consignment size	703,996	62,278	56,078			
Consignments infested with a quarantine pest	6	0	60			
Expected number of consignments in the model[a]	600	125	700			
Estimated proportion of infestation p_{ij}[b]	4.97E-05	1.05E-07	3.65E-04			
B						
Consignments imported during 1998-2001	1,008		7,255			818
Average consignment size	7,743		34,318			30,023
Found infested with a quarantine pest	8		106			0
Expected number of consignments in the model[a]	150		1,815			210
Estimated proportion of infestation p_{ij}[b]	7.43E-05		2.02E-04			1.22E-7
C						
Consignments imported during 1998-2001	606			586	6,354	
Average consignment size	1,013,192			5,477	23,109	
Consignments infested with a quarantine pest	1			0	2	
Expected number of consignments in the model[a]	155			150	1,560	
Estimated proportion of infestation p_{ij}[b]	1.20E-06			9.33E-07	3.54E-05	

* 9 pathways (3 ornamental species coming from 3 countries) are represented
[a] based on 2001 data
[b] own estimation

The rejection rate of consignments is not sufficient to deduce the true proportion of infestation p_{ij} of a given pathway. Reliable data of the proportion of infestation can be obtained only when the exact number of infested plants in every consignment found infested is counted[3]. Unfortunately, such data were not available for our model. To estimate the proportion of infestation we used the following approaches. If no consignments of the ijth pathway were rejected during import inspection, p_{ij} was estimated using the upper 95 % confidence limit using formula $0.95 = 1 - (1 - p_{ij})^{v_{ij}}$ from Couey and Chew (1986), where v_{ij} is the total number of plants imported along the ijth pathway in 1998-2001.

When the number of rejected consignments was greater than zero, we used equation (5) to estimate p_{ij}. Parameters u_{ij} and v_{ij} in this equation were taken from data for 1998-2001 shown in Table 2. Parameter p_{inf} in the same equation was given by the mean of Triang (0.5 %, 10 %, 20 %) distribution where parameters represent the minimum, most likely and maximum values, respectively. This distribution is assumed to approximate the variation in the actual proportion of infestation in consignments found infested with a quarantine pest. Although difficult to justify empirically, both the distribution and chosen parameters were based on a number of considerations. The lower bound was set on the presumption that because infestation was detected, the infestation rate was at least 0.5 %, i.e. the level at which the quarantine inspection can detect infestation with reasonable confidence. The only evidence for the most likely value comes from Frey (1993). Examining the infestation rate of ornamental cuttings imported into Switzerland he found an average sample infestation rate of approximately 13 %, ranging from 2.3 % for Dianthus and 8 % for Impatiens to 15 % for Chrysanthemum. With large uncertainty we set the most likely value at 10 %. The choice of the upper bound was based on the idea that the phytosanitary quality of imported commodities is currently high. This is because: 1) the exporting countries' inspecting Agency would detect sufficiently low infestations, and 2) even if an infestation is missed by the export quarantine, the commodities are chilled during transportation and infestation rate at the time of import inspection is unlikely to exceed seeming reasonable 20 %. The estimated proportions of infestation (p_{ij}'s) are shown in Table 2.

RESULTS

Before discussing the results of the model, it is useful to estimate expected pest risks in the absence of import inspection. This will allow seeing the effect of import inspection better. Recall that in our model quarantine risk is measured as the expected number of infested plants entering the importing country. Straightforward application of equation (4) yields the required estimate. Thus, the expected number of infested plants in the absence of inspection is calculated as the product of the expected volume of imported plants and the estimated proportion of infestation associated with the given pathway. Parameter α is equal to unity in this case to reflect the absence of import inspection. The resulting risk estimates for different pathways are presented in Table 3 below.

Table 3. *Expected number of infested plants per pathway*[*]

Country	Ornamental species					
	Chrysanthemum	Rose	Dianthus	Yucca	Dracaena	Impatiens
A	23,905	<1	21,172			
B	265		23,112			<1
C	177			<1	1,289	

[*]Calculated as the summed product of p_{ij} (Table 2) and the average consignment size in each of consignment size categories

Table 3 shows that the largest number of infested plants is expected from Dianthus pathways, reflecting relatively high proportions of infestation and volumes (especially in terms of number of consignments). Large numbers of infested plants can be also expected from ChrysanthemumA pathway, reflecting mainly the large volume of incoming plants along this pathway. As can be expected, pathways with higher proportions of infestation and large volumes of import represent the largest quarantine threat. Pathways with estimated (very) low p_{ij} thus represent a lower quarantine risk. The total number of infested plants expected from all pathways is about 69,872. The average proportion of infestation is approximately equal to 0.0001 (69,872 / 671 million).

To obtain a plausible value for the constraint B we then ran the model for the situation that is assumed to reflect current inspection practices. Here, the Agency applies the same inspection treatment to all pathways. The inspection length is fixed at 30 minutes with an error level of approximately 5 % (see Table 1). The resulting costs of inspections are obtained by multiplying the corresponding inspection tariff (i.e. 83.28 euros) with the total number of consignments imported along all 9 pathways. The costs per pathway were defined only by the number of consignments to be imported along a given pathway. The resulting total inspection cost amounted to 455,125 euros. The expected number of infested plants after application of such a uniform inspection rule is approximately equal to 4,010. The efficacy of quarantine inspection is thus about 94.3 % (1 - 4,010 / 69,872).

It is the total inspection costs obtained in the model above (i.e. 455,125 euros) that were used as a constraint in the main optimization model. The model should thus allocate these funds freely to find the solution in which the expected number of infested plants imported into the country is minimal. Table 4 presents the results of the budget allocation between the pathways in the model.

Table 4. *Budget allocation per pathway, after minimizing risk (1000 euros)*

Country	Ornamental species					
	Chrysanthemum	Rose	Dianthus	Yucca	Dracaena	Impatiens
A	85,67	–	86,30			
B	7,32		169,22			–
C	12,50			–	94,11	

In Table 4, the sum of all pathway budgets equals the value of the constraint, i.e. 455,125. The budget is thus fully used. The allocation of budget to pathways is, however, very different. First, note that no budget at all is allocated for inspection of Rose, Yucca and Impatiens. This is consistent with the very small quarantine risks that they pose (see Table 3). Among pathways with a positive budget allocation, the largest shares of total budget are allocated for inspection of DianthusB and Dracaena. The DianthusB pathway received a large allocation because of both a high number of infested plants expected and a large number of imported consignments. The large absolute inspection costs allocated for Dracaena pathway are explained mainly by the large expected number of imported consignments; the quarantine threat posed by Dracaena is much lower than, for example, by ChrysanthemumA (see Table 3). In general, the results of budget allocation presented in Table 4 are consistent with numbers presented in Table 3. Pathways with larger expected number of infested plants *ceteris paribus* receive larger budget allocation. To see *how* pathways budgets are allocated, let us inspect Figure 1.

Figure 1 shows the distribution of the inspection lengths for a given pathway for consignments of different sizes within the pathway. Figure 1 indicates that budget as a function of inspection time is allocated differently not only across pathways, but also across different consignment size categories within pathways. The general trend is that larger consignments receive lengthier inspection treatment than smaller ones. Furthermore, pathways with larger expected number of infested plants *ceteris paribus* are inspected with more time. Compare again results for DianthusB and Dracaena pathways. The consignments coming along the former pathway should be inspected with more time than consignments coming along the latter. This finding reflects the difference in quarantine risks between these two pathways and supports an earlier argument that Dracaena received large absolute budget allocation mainly because of the large number of imported consignments.

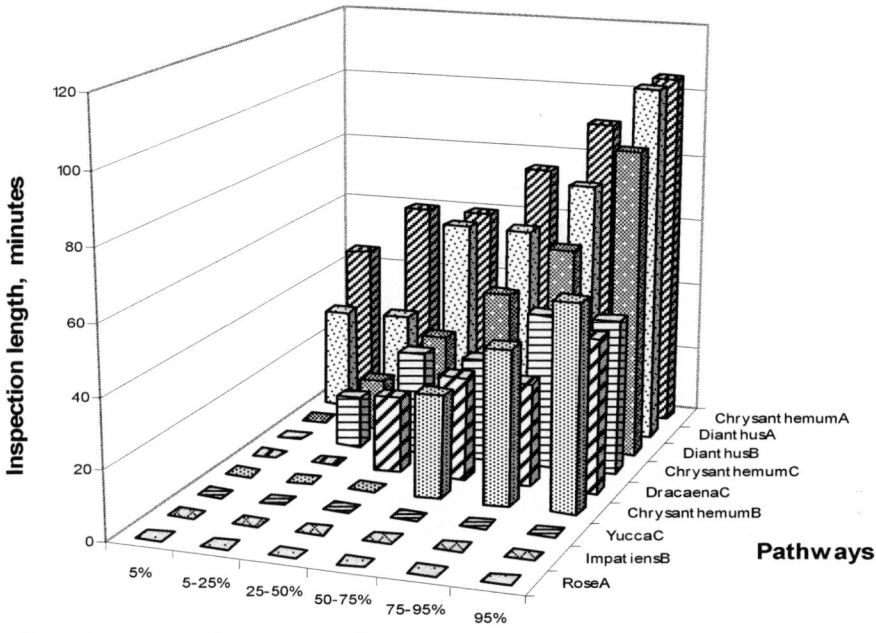

Consignment size percentiles

Figure 1. *Distribution of inspection times across pathways and sizes groups*

The expected number of imported infested plants in this model is equal to 380, suggesting that the Agency may reduce the initial risk by 99.4 %. This is due to allocation of larger budgets and longer inspection times for *a priori* more risky pathways. In fact, there is a redistribution of the common resources towards riskier pathways at the expense of pathways with comparatively lower risks. This explains why the reduction in the expected risk in this model is higher compared to the model in which all pathways are inspected with equal time and budget per inspection. On the other hand, some pathways (Rose, Yucca and Impatiens) remain completely uninspected implying that the Agency should bear the risk that some infested plants might be imported along these pathways.

It is worthwhile noting that obtained results remain stable when there is a change in the quarantine budget. An increase (decrease) in the total budget leads to an increase (decrease) in the average time of inspection of a pathway. The direction of budget distribution also remains consistent with observed trends: more risky pathways and larger consignments receive proportionally higher budgets. Another important result is related to the shadow price of the budget constraint. Recall from the theoretical model that the shadow price indicates the change of objective value had the constraint been changed by one euro. The shadow price in the model was equal to -0.0032, implying that the 312.5 euro increase in the total budget would lead to approximately 1 unit decrease in the expected number of infested plants. A 50 % increase (decrease) in the total budget resulted in shadow prices equal to -0.00032 (-0.0198). These results are in line with the premise that import inspection has high marginal efficacy with low budgets and low efficacy with high budgets (because it is more difficult to detect a marginal infested plant).

DISCUSSION

In this paper we presented a model of optimal allocation of budget resources to minimize import quarantine risks. The theoretical model implies that the available resources should be allocated so that the marginal pest risks are equalized across import pathways. The results of the empirical model suggest that pathways with larger expected risks *ceteris paribus* should receive a larger share of the budget and longer inspection treatment. Within pathways, larger consignments must be inspected more intensively than smaller ones. This finding reflects the implicit assumption that for a fixed proportion of infestation, larger consignments have more infested plants, and thus require more thorough inspection treatment (assuming that the probability of detecting a pest does not depend on the consignment size). The model output also suggests that some pathways with *a priori* low risks may remain completely uninspected. This finding is consistent with Horan et al. (2002, p. 1309), who noted that it is optimal to devote more resources to confront (quarantine) events that are considered more likely and to allocate few or no resources to confronting events that are considered less likely. Yet, it is obvious that the Agency should be prepared to bear some quarantine risks in this case (due to no inspections of certain pathways).

The main message from these results is that, with limited resources, the inspection of all risky pathways may not be optimal (let alone feasible). For quarantine policy making, this implies that the Agency should focus on, *ceteris paribus*, riskier pathways and leave other pathways uninspected or inspected with lower effort. Presumably, this is the current practice in many countries worldwide. A possible solution to alleviate the quarantine risks remaining along unchecked pathways would be for the Agency to rely on self-protection efforts of importers of risky commodities (or other interested stakeholders).

Some reservations related to the model setup and assumptions should be mentioned. The first reservation is related with data. Quantitative data related to quarantine risks are generally scarce (Gray et al. 1998) and the proportions of infestation are very hard to estimate at the low levels that are prevalent. However, the actual application of the model developed in this paper crucially depends on the availability and quality of the quantitative estimates. The procedure to estimate the proportion of infestation – a key factor influencing the optimal allocation of resources among different pathways – in the current work was indirect, implying that the estimates of p_{ij} may be biased. This bias may be in part due to a triangular distribution used to estimate the proportion of infestation in rejected consignments. Conceivably, this distribution gives only a limited approximation of the true proportion of infestation. Given that the exact computation of actually infested plants is almost infeasible, other non-parametric distributions with more parameters (for example, discrete) could be used as possible alternatives. Data on parameters in these distributions may come from experts.

The discussion in the previous paragraph underscores the importance of the proper account of uncertainty in estimating quarantine risks associated with different pathways. Another important characteristic that the model fails to address is the variability in quarantine risks (Gray et al. 1998). The model found the optimal solution based on the premise that the proportion of infestation of a given pathway is fixed. Specifically, it was expressed as the mean of the probability distribution $f(p_{ij})$ of the proportion of infestation. As a result, in the model every consignment is assumed to carry a positive number of infested plants, which is somewhat counterintuitive. In reality one would expect a very significant variation in the p_{ij} within the pathway, e.g., due to stochastic fluctuations or to variations in the quality of plants imported from different producers in the exporting country. This variation most probably takes the form that some of the consignments are completely free from quarantine organisms (after all, most consignments successfully pass import inspection) and others are infested with varying extent. A more realistic model should take this issue into account.

These shortcomings suggest clear avenues for improvement of the presented empirical model. Overall, we believe that the presented model is a useful step towards development of more effective quarantine inspection policy.

ACKNOWLEDGEMENTS

The Dutch Plant Protection Service (PD) is acknowledged for providing data on import inspections. The authors thank Jan Schans of the PD for stimulating discussion and very helpful comments on the paper. Comments of Christien Ondersteijn at the earlier stage of this work are gratefully appreciated. Authors also thank Paul Berentsen and Annemarie Breukers for critical and very helpful comments on the manuscript. Any remaining errors are ours.

NOTES

[1] There is a third option: to minimize risks for specific pests (Bigsby 2001); we ruled this possibility out by assuming non-pest-specific risks

[2] We use the term 'quarantine' throughout the remainder of the paper to emphasize that the pest that caused the rejection of a particular consignment was not tolerated by the importing country. In reality, consignments in the database were rejected due to both quarantine and non-quarantine pests; however, for the purposes of the numerical model we consider all cases of rejections as due to quarantine pests. This is consistent with the set-up of the model, in which Agency considers all pests as equally damaging. For official definition of the quarantine pest see FAO (2006)

[3] This is the approach adopted by e.g. Roberts et al. (1998) and Wearing et al. (2001) in the quantitative risk assessments of, respectively, fire-blight and codling-moth introductions via trade in fruits

REFERENCES

Barbier, E.B., 2001. A note on the economics of biological invasions. *Ecological Economics,* 39 (2), 197-202.
Batabyal, A.A. and Beladi, H., in press. International trade and biological invasions: a queuing theoretic analysis of prevention problem. *European Journal of Operational Research.*

Bigsby, H.R., 2001. The 'appropriate level of protection': a New Zealand perspective. *In:* Anderson, K., McRae, C. and Wilson, D. eds. *The economics of quarantine and the SPS agreement.* Centre for International Economic Studies, Adelaide, 141-163.

Chiang, A.C., 1984. *Fundamental methods of mathematical economics.* 3rd edn. McGraw-Hill, Auckland.

Couey, H.M. and Chew, V., 1986. Confidence limits and sample size in quarantine research. *Journal of Economic Entomology,* 79 (4), 887-890.

EPPO, 2005. *Sampling of consignments for visual phytosanitary inspection: draft standard.* European and Mediterranean Plant Protection Organization. Available: [http://www.eppo.org/News&Events/05-11611%20PM3%20sampling.pdf] (July 2005).

Everett, R.A., 2000. Patterns and pathways of biological invasions. *Trends in Ecology and Evolution,* 15 (5), 177-178.

FAO, 2006. *International standards for phytosanitary measures: glossary of phytosanitary terms.* Secretariat of the International Plant Protection Convention FAO, Rome. ISPM no. 05. [https://www.ippc.int/servlet/BinaryDownloaderServlet/133607_ISPM05_2006_E.pdf?filename=115 1504714760_ISPM05_2006_E.pdf&refID=133607]

Frey, J.E., 1993. The analysis of arthropod pest movement through trade in ornamental plants. *In:* Ebbels, D. ed. *Plant health and the European single market: proceedings of a symposium organised by the British Crop Protection Council, the Association of Applied Biologists and the British Society of Plant Pathology, and held at the University of Reading, 30th March-1st April 1993.* BCPC, Farnham, 157-165. BCPC Monograph no. 54.

Gray, G.M., Allen, J.C., Burmaster, D.E., et al., 1998. Principles for conduct of pest risk analyses: report of an expert workshop. *Risk Analysis,* 18 (6), 773-780.

Horan, R.D., Perrings, C., Lupi, F., et al., 2002. Biological pollution prevention strategies under ignorance: the case of invasive species. *American Journal of Agricultural Economics,* 84 (5), 1303-1310.

Kuno, E., 1991. Verifying zero-infestation in pest control: a simple sequential test based on the succession of zero-samples. *Researches on Population Ecology,* 33 (1), 29-32.

Ministry of Agriculture and Forestry, 2006. *Biosecurity New Zealand, standard 155.02.06, importation of nursery stock.* Ministry of Agriculture and Forestry, Biosecurity New Zealand, Wellington. [http://www.biosecurity.govt.nz/imports/plants/standards/155-02-06.pdf]

National Research Council, 2002. *Predicting invasions of nonindigenous plants and plant pests.* National Academy Press, Washington. [http://www.nap.edu/books/0309082641/html/]

Olson, L.J. and Roy, S., 2002. The economics of controlling a stochastic biological invasion. *American Journal of Agricultural Economics,* 84 (5), 1311-1316.

Plantenziektenkundige Dienst, 2005. *Tarievenoverzicht Plantenziektenkundige Dienst.* Ministerie van Landbouw, Natuur en Voedselkwaliteit, Plantenziektenkundige Dienst. Available: [http://www9.minlnv.nl/pls/portal30/docs/FOLDER/MINLNV/LNV/UITVOERING/UD_PD/TARIE VEN/TARIEVENOVERZICHT_PD_OKT05.PDF] (July 2006).

Roberts, R.G., Hale, C.N., Van der Zwet, T., et al., 1998. The potential for spread of *Erwinia amylovora* and fire blight via commercial apple fruit: a critical review and risk assessment. *Crop Protection,* 17 (1), 19-28.

Saphores, J.-D.M. and Shogren, J.F., 2005. Managing exotic pests under uncertainty: optimal control actions and bioeconomic investigations. *Ecological Economics,* 52 (3), 327-339.

Wearing, C.H., Hansen, J.D., Whyte, C., et al., 2001. The potential for spread of codling moth *(Lepidoptera:Tortricidae)* via commercial sweet cherry fruit; a critical review and risk assessment. *Crop Protection,* 20 (6), 465-488.

QUANTIFYING RISKS AND ECONOMIC EFFECTS USING SPATIAL MODELS

CHAPTER 5

EVALUATING THE COST-EFFECTIVENESS OF BROWN-ROT CONTROL STRATEGIES

Development of a bio-economic model of brown-rot prevalence in the Dutch potato production chain

ANNEMARIE BREUKERS, MONIQUE MOURITS, WOPKE VAN DER WERF, DIRK L. KETTENIS AND ALFONS OUDE LANSINK

Business Economics, Social Sciences Group, Wageningen University, Hollandseweg 1, 6706 KN Wageningen, The Netherlands. E-mail: annemarie.breukers@wur.nl

Abstract. Quarantine diseases comprise a distinct class of plant diseases. In contrast to other plant diseases, direct losses through crop damage are often limited. Yet, quarantine diseases may have serious economic consequences for a country as they threaten the country's export of affected crops. In particular for this category of diseases, it is important to design a control strategy that is optimal from an epidemiological as well as an economic point of view. This chapter presents the development of a bio-economic model to evaluate control strategies in terms of their cost-effectiveness, specified for brown rot in the Dutch potato production chain.

The conceptual model consists of two modules: an epidemiological module, which is a stochastic, spatially explicit simulation model that simulates the spread of potato brown rot over all potato-growing farms and fields in the Netherlands, and an economic module, which calculates the total costs of brown-rot prevalence, based on the results of the epidemiological model.

The model is applied for two brown-rot policy scenarios, which differ in sampling frequency of harvested potato lots but are otherwise similar. The two scenarios are compared with respect to their effectiveness and efficiency. Concerning the costs of controlling brown rot, a low monitoring level appears to be more cost-efficient; however, when including expected export consequences, a high monitoring level may be preferable.

The model presented here strongly facilitates the development of an optimal control strategy as it provides insight into the effectiveness of brown-rot control strategies in relation to their costs. Moreover, the introduced modelling concept can be a useful tool in analysing the epidemiological and economic effects of other (quarantine) diseases.

Keywords: *Ralstonia solanacearum*; epidemiology; economics; quarantine disease; simulation model

A.G.J.M. Oude Lansink (ed.), New Approaches to the Economics of Plant Health, 57–70.
© 2007 *Springer.*

INTRODUCTION

Quarantine pests and diseases comprise a distinct class of plant diseases (Heesterbeek and Zadoks 1987). A quarantine status is assigned to diseases that are not yet present, or present but not yet established in a region and can potentially cause serious economic damage in this region (IPPC 1999). The emergence of a quarantine disease in a country involves the imposition of a national control policy, which aims at eradication of the disease and prevention of new introductions. Such policy often includes a set of measures that bring along high costs for both the government and the stakeholders of the production chain to which the measures apply, whereas the disease in itself may cause only limited or even no damage at all to the host crop. The costs of controlling the disease may thus far exceed the direct benefits of avoiding yield losses, whereas long-term benefits of the control policy are unclear.

An example of a quarantine disease with potentially high costs is brown rot in the Dutch potato production chain. The disease is caused by the bacterium species *Ralstonia solanacearum* race 3, biovar 2, which is pathogenic on, e.g., potato, tomato and several solanaceous weeds. In warm and humid growing areas, the disease can be very destructive. Outbreaks have been reported in many European countries, and, within the EU, brown rot has obtained a quarantine status (Elphinstone 2005). In the Netherlands, climatic conditions are less favourable for brown-rot population growth, and infections generally remain symptomless. Nevertheless, its presence may have serious economic consequences for the Dutch potato production chain. The risk of establishment of brown rot as an endemic disease threatens the Dutch market share of seed potatoes, which comprise an important export product of the Netherlands (Van Vaals and Rijkse 2001). To avoid economic losses resulting from reduced export, the government has imposed a costly control policy aimed at eradication of the disease from the chain.

The number of detected brown-rot infections has strongly decreased since the first outbreak in 1995. However, the intended eradication of the disease from the chain has not been achieved. The set of brown-rot measures that are currently in force are generally based on practical experiences with brown rot so far; theoretical evidence of the efficacy of a measure is often lacking. Whereas the major risk factors responsible for brown-rot prevalence dispersal are known, quantitative knowledge about the risk-reducing effects of control measures is still poor. Moreover, although the currently implied control policy seems to be effective in reducing the number of outbreaks, insight into its cost-effectiveness is lacking. In other words, possibly the same result could be achieved at reduced costs, or with a similar budget the effectiveness could be increased.

In this chapter, we present a bio-economic model to evaluate brown-rot control strategies in terms of their cost-effectiveness. The model quantifies the total costs of brown-rot prevalence, based on the simulation of brown-rot dynamics for a specific control scenario. It provides insight into the relative importance of risk factors and the efficacy of control measures in relation to their implementation costs. Thereby, the model facilitates the design of an optimal control policy. In the next section, the conceptual framework of the bio-economic model is explained. In the third section, the conceptual model is applied for the above-described case of brown-rot prevalence in the Dutch potato production chain. Finally, wider implications of this model and future perspectives are discussed.

CONCEPTUAL FRAMEWORK

The bio-economic model consists of an epidemiological module and an economic module. The epidemiological module is a stochastic, spatially explicit simulation model that simulates the spread of potato brown rot over all potato-growing farms and fields in the Netherlands over a chosen time frame. It can be run independently from the economic module to evaluate the effectiveness of control strategies. The economic module calculates the total costs of brown-rot prevalence based on the results of the epidemiological model.

Figure 1 shows a schematic representation of the bio-economic model. For a given control strategy and time frame, the epidemiological model is run to simulate brown-rot dynamics over the indicated period. As this model is stochastic, numerous replications should be run in order to obtain a representative picture of brown-rot incidence under a chosen control strategy, parameterization or set of conditions. The results of each replication are recorded in output files, which serve as input for the economic module. This module subsequently calculates the yearly costs related to brown-rot prevalence and control. The economic module does not contain any stochastic elements. However, the output of the epidemiological model contains results of all replications, so the costs can be calculated for each year in each replication, resulting in a distribution of yearly costs. In the following two subsections, the epidemiological and economic modules are explained in more detail.

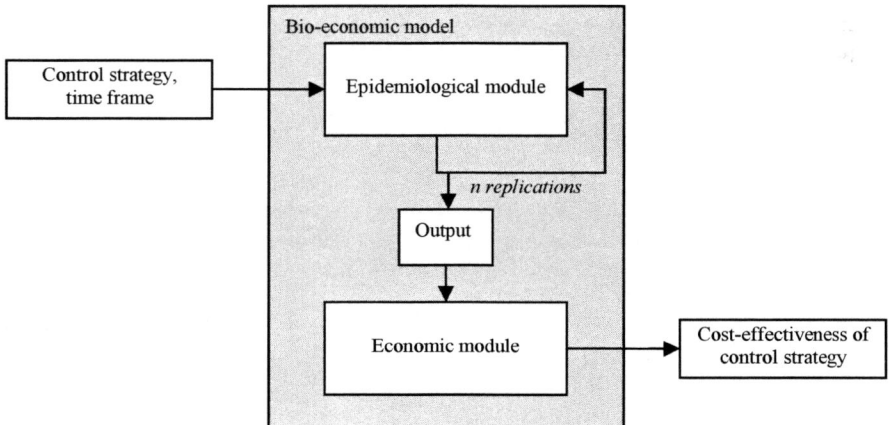

Figure 1. Conceptual framework of the bio-economic model

Epidemiological module

Simulation of disease dynamics at the level of the plant production chain is still rather uncommon (Breukers et al. in press). As local circumstances of, and contacts between different infectious units are for a large part determined by human-driven processes, the epidemics of a disease are not solely determined by the pathogen's biological characteristics, and conventional epidemiological models are not appropriate anymore. In our attempt to simulate disease dynamics in the potato production chain, we used an individual-based modelling technique, which originates from the field of ecology. The essence of this technique is that it acknowledges and explicitly represents the principle that each individual is unique in its characteristics and interactions with other individuals. An individual-based model (IBM) traditionally defines the individual organism as the logical basic modelling unit instead of using aggregated-state variables to describe population dynamics (Huston et al. 1988; Grimm 1999). In our application, the modelled 'individual' is the potato lot, which is the commercial production entity of the potato production chain.

The modelled potato production chain starts with the multiplication of high-quality seed potatoes. After several years of multiplication, these seed potatoes are grown as ware or starch potatoes, which are eventually transported to retail or industry; this is where the modelled production chain ends. The production chain contains three possible pathways for brown-rot infection: primary infection, horizontal transmission and vertical transmission. Primary infection may occur as a consequence of irrigation of a potato lot with surface water; large parts of the Dutch surface water are contaminated with brown-rot bacteria as a result of the common presence of the wild host plant, woody nightshade (*Solanum dulcamara*). Horizontal transmission means infection of a healthy potato lot where the source is another infected lot, and can for instance be caused by poorly separated storage of potatoes or planting or harvesting a lot with contaminated machinery. Vertical transmission, also referred to as infection through clonal relationships, implies transmission of the disease from 'parent' to 'offspring', i.e. from one generation to the next.

The individual farms, fields and potato lots in the above-described system are represented in the model by so-called objects. Each object is described by a set of variables, or 'attributes'. The values of these attributes describe the state of each object and make each object unique. Examples of attributes are the size of a field object, the potato acreage of a farm object, and the detection status of a potato lot object. To include spatial explicitness, farm and field objects have attributes indicating their geographical location. Objects can be linked with other objects to represent certain relations between individuals. For example, an infected lot is linked to the field on which it is grown, and a farm is linked to the fields it has in use.

During a simulation run, the model separately keeps track of the dynamics of all objects included in the simulation, taking into account their unique properties and interactions. Each year of simulation consists of one production cycle, which in turn consists of a number of processes, such as planting, irrigation, testing and storage. These processes may go together with 'events', which affect the attribute values and thus the state of objects. Examples of events are 'primary infection of a lot' during irrigation, or 'detection of a lot' when an infected lot is tested for brown rot. The occurrence of an event is determined stochastically and depends on one or more parameters.

Information about the state of the system is recorded and stored in output files. At the end of a simulation, the output files contain detailed information about the number and characteristics of infected lots and the farms and fields on which they were grown. The results may strongly vary across years and replications, which is partly due to the stochasticity of most events related to infection and detection of potato lots. Another part of the variation is explained by the dynamic nature of brown rot; the number of infections in one year is partly determined by the incidence of infections in previous years. The output files of the epidemiological model serve as input for the economic module. As the output includes the variation in brown-rot prevalence, the total costs of brown rot can be presented by a frequency distribution, which allows for calculating both the average costs and the likelihood of extremely high or low costs.

The structure of an IBM is similar to the structure of an object-oriented program (Acock and Reddy 1997); therefore, the epidemiological model described here is implemented in the object-oriented programming language C++. A detailed description of the modelled system and conceptual epidemiological model is given in Breukers et al. (2005).

Economic module

The economic module calculates the total costs of brown rot for a given control policy. It distinguishes three categories: (1) structural costs, which are incurred yearly as a result of preventive measures; (2) incidental costs, which only occur in case of brown-rot detection and result from eradication measures; (3) export losses, which arise as a consequence of failure to reduce brown-rot incidence to an acceptable level. One could also discriminate between direct costs, which are clearly attributable to an outbreak, and indirect costs, which are independent of the brown-rot incidence and are incurred also by individuals that are not involved in an outbreak. Incidental costs correspond with direct costs, while structural costs and export losses can alternatively be referred to as indirect costs. Costs are incurred by stakeholders in the potato production chain, such as farmers and trading companies, as well as the governmental authorities. Possible economic consequences for other sectors, for instance the transport sector or crop protection companies, are not included in the analysis. Below, each category of costs is described in more detail and its method of calculation is explained. An overview of all cost items is given in Table 1.

Structural costs

Structural costs are incurred yearly and are more or less stable over time for a specific control strategy. They result from preventive measures against introduction and dispersal of brown rot, such as monitoring of potato lots and other potential brown-rot sources. Monitoring is done by the authorities, but growers are (partly) charged for the sampling costs of potato lots. Another preventive measure is the prohibition of surface-water use in areas where this water may contain brown-rot bacteria. The ban on the use of surface water causes losses for potato growers and trading companies due to yield reduction and quality disorders caused by common scab. These losses of brown rot only occur in areas where alternatives to surface water (e.g., groundwater) are not available.

The total structural costs are calculated by summing up all components of structural costs.

Incidental costs

Incidental costs are directly related to eradication measures following detection of an infected lot, such as destruction of a detected lot and tracing of lots that might have been in contact with this lot. Furthermore, lots that are not found to be infected in a test but nevertheless are highly suspected because of clonal or spatial relationships with an infected lot, are defined as 'probably infected' and downgraded. Such lots cannot be replanted and can be marketed only under strict conditions, often for low prices. The losses for a farm resulting from destruction and downgrading are considerable. Seed-potato growers multiplying their own seed potatoes for several successive years face even higher losses, as they have the costs of buying new planting material. Furthermore, affected growers may earn lower revenue per hectare for one or more years after detection as a result of restrictions on the crops that can be grown on the farm and field on which a detected lot was grown. Labour costs at farm level are not included; it is assumed that a farmer will not hire (extra) labour and that the opportunity costs of the farmer's own labour are zero.

The incidental costs for affected potato growers are calculated by partial budgeting. This technique quantifies the economic consequences resulting from a change at the farm level (i.e. detection of an infected lot). Four components are accounted for: extra costs, returns forgone, reduced costs, and additional returns (Rushton et al. 1999; Dijkhuizen and Morris 1997). The tracing costs are determined by the number of extra samples that must be taken, which is given by the epidemiological model. All costs made at farm level and the costs of tracing are subsequently summed up to obtain the total incidental costs.

Export losses

As a result of inability to achieve a low and stable level of brown-rot incidence, the potato sector, or even the national economy, may experience negative consequences from the presence of brown rot. Prolonged presence of brown rot in the Dutch potato production chain will harm the image of the sector and its products in importing countries, which has a negative effect on export volume and consequently on the potato price. In particular, export of seed potatoes will be reduced, as these potatoes are replanted and thus comprise a high risk of introducing brown rot in the potato production chain of the importing country. Lower prices may in the long run result in a reduction in national potato production, as growers will replace part of their potato acreage by other, more profitable crops.

The relation between brown-rot incidence and the level of export consequences is difficult to quantify. Therefore, the losses due to export consequences are qualitatively assessed based on the yearly number of exported infected lots, which is an output of the epidemiological model.

Table 1. Overview of direct and indirect costs per category and underlying measure or causal factor. Stakeholders responsible for the costs are indicated by capitals: F = farmer; T = trading company, retail and industry; G = government; S = entire sector

Measure or cause	Costs	Payer(s)
Structural costs		
sampling of potato lots for brown-rot prevalence	costs of sampling	F, G
sampling of other potential sources for brown-rot prevalence	costs of sampling	G
prohibition on use of surface water	yield and quality loss	F, T
Incidental costs		
destruction of detected potato lots	loss of revenue	F, T
	replacement of propagation material	F
	destruction costs	F
downgrading of probably infected potato lots	loss of revenue	F, T
	replacement of propagation material	F
tracing of potato lots related to detected lots	costs of sampling	G
increased sampling intensity on affected farms	costs of sampling	G
monitoring on affected farms	labour costs	G
prohibition of seed potato production for 1 or 2 years	production of less profitable crops / smaller acreage	F
quarantine status on field	cultivation of less profitable crops / rent of other field	F
Export losses		
loss of international market share	effects on (seed) potato price	S
	effects on (seed) potato acreage	S

APPLICATION OF THE BIO-ECONOMIC MODEL

In this section, simulation results are presented for two different brown-rot policy scenarios. The first scenario represents the policy applied in the Netherlands until 2004. This scenario includes an intensive monitoring strategy: all seed lots grown according to the standards of the Dutch General Inspection Service (Nederlandse Algemene Keuringsdienst, NAK) are sampled during or shortly after harvest at an intensity of 200 tubers per 25 tonnes. Ware and starch potatoes are sampled randomly at a frequency of approximately 7 % (one sample per lot). If a lot is found to be infected, it is destroyed, and quarantine measures are enforced on the farm that owned the lot and the field on which it was grown. All lots that are clonally related to the detected lot or may have been in contact with it are traced and tested for brown rot. Lots that are not found infected but that are nevertheless strongly suspected of being infected, as for example all other lots grown on the infected farm, are defined 'probably infected' and are downgraded to ware or starch potatoes, depending on their variety. The second scenario differs from the first in that only 10 % of all seed lots are randomly sampled, as compared to 100 % in the first scenario. The two scenarios will further be referred to as the 'base' and 'reduced sampling' scenario, respectively.

The system represented by the model comprises the Dutch potato production chain, including all potato-growing farms and arable fields in the Netherlands. Data on these farms and fields were obtained from the Dutch Agricultural Research Institute (Landbouw Economisch Instituut, LEI). Epidemiological and sector parameters required for the epidemiological model are derived from elicitation of experts (Breukers et al. in press). Economic data were provided by the PPO (Praktijkonderzoek Plant & Omgeving) research unit Arable Farming (Dekkers 2001), the brown-rot insurance company PotatoPol, and various stakeholders of the Dutch potato production chain. Results presented in this section are based on 100 replications for each scenario, each replication covering a period of 15 years. At this number of replications, the change in the running mean of the average yearly number of infections per simulation is less than 0.5 % and the computer time required for a simulation is still acceptable. The results are based on years 6-15 of the simulation runs to exclude initial transitory effects.

Brown rot dynamics

Figure 2 shows the normal and cumulative distribution of the yearly number of infected lots over all years of all replications (n=1500) for both policy scenarios. For the base scenario, the yearly number of infections varies between zero and approximately 80, but is equal to or lower than 10 in more than 50 % of the observations. This result corresponds with the typical irregular pattern of brown-rot dynamics observed in practice (Hendriks and Höfte 2004). Occasional outbreaks of brown rot can be caused by weather circumstances such as a conducive (i.e. warm and humid) summer, which is a stochastic model variable. Outbreaks may also occur when one or more infections in seed lots in a certain year are not detected; these seed lots are split in approximately 5 daughter lots on average, which are replanted in the following year.

Comparing the base scenario with the reduced-sampling scenario shows that reducing the sampling frequency mainly affects the median and variation in yearly number of infection, while leaving the mode (i.e. the most frequently observed number of infections per year), rather unaffected. Outbreaks of over 100 cases per year are occasionally observed, and the median almost doubles to 19 infections per year. The explanation for this difference is that, at a reduced sampling frequency, the detection efficiency of infected seed lots decreases. Consequently, infected seed lots have a higher probability of remaining in the production chain and being split into daughter lots for several years, occasionally leading to high numbers of infected lots.

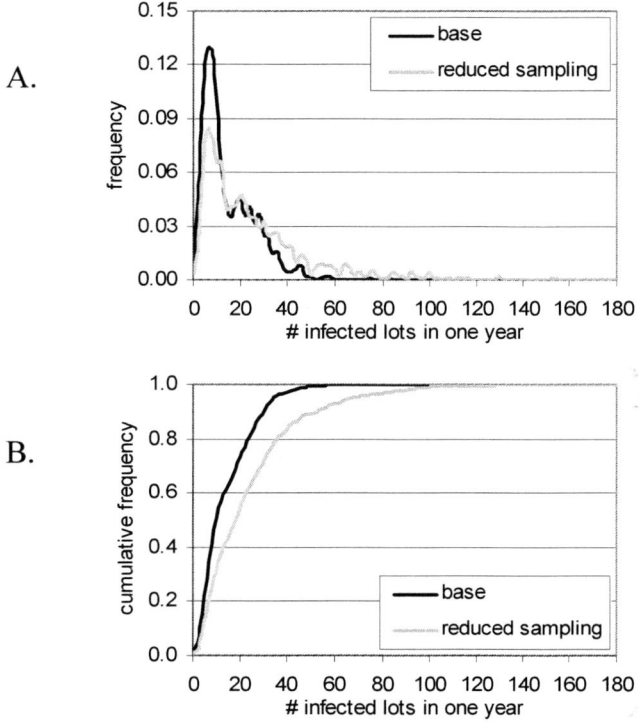

Figure 2. *Normal (a) and cumulative probability density diagram (b) of the observed yearly number of infections over 100 replications of 15 years*

Structural and incidental costs

The structural and incidental costs are presented at national level; no specification is given of the costs for different groups of stakeholders in the potato production chain. The only structural costs calculated in this chapter include the yearly sampling costs of potato lots. Other structural costs will be more or less the same for the two scenarios evaluated here. Structural costs are assumed constant over time and are calculated based on a fixed potato acreage and average lot size and yield per hectare. Incidental costs are calculated separately for each year in each simulation run and include all structural and incidental costs. The economic module that is currently available is preliminary and does not include the costs related to the quarantine status on fields yet, as these are still under investigation. The module assumes that, in case seed-potato production is prohibited on a farm, this farm will produce ware potatoes or starch potatoes instead.

The total production of seed potatoes in the Netherlands is approximately 1.4 mln tonnes on average. Under the base scenario, this quantity results in approximately 55,000 samples, whereas under the alternative scenario only 10 % of these samples are taken. The costs of sampling ware and starch lots are the same under both scenarios. In total, the yearly sampling costs comprise on average almost 4.5 million euros per year for the base scenario, and 0.7 million euros for the reduced-sampling scenario. The variation in sampling costs between years is negligible, as the total potato acreage planted – and thus the number of samples taken – remains stable over time.

In contrast to the structural costs, the incidental costs can vary strongly across years, as they are directly related to the number of detected lots in the current and previous year. Figure 3a shows the distribution of the incidental costs per year for both scenarios. For both scenarios, the incidental costs remain far below one million euros in most years. The median for the base scenario lies at 0.19 million euros per year, as compared to 0.11 million euros per year for the reduced-sampling scenario. Occasionally, the costs for the reduced-sampling scenario are much higher than for the base scenario. This result reflects the high variation in the number of brown-rot cases per year that was already observed from the epidemiological output of the reduced-sampling scenario (Figures 2A and 2B). The two incidental costs lines intersect; consequently, there is no first-order stochastic dominance, i.e. it is not possible to select one scenario which is always 'better' than the other. Risk-averse decision makers will in the first place try to minimize the risk of incurring extremely high costs, whereas risk-neutral decision makers attach most importance to average costs. The average yearly incidental costs of the base and reduced-sampling scenario are 0.31 and 0.41 million euros, respectively, so the base scenario will be preferred over the reduced-sampling scenario, irrespective of the risk attitude of the policy makers.

When summing up the structural and incidental costs (Figure 3B), it turns out that the reduced-sampling scenario is dominant over the base scenario to the first order; i.e. the curve of the reduced-sampling scenario is always preferred over the base scenario. Even in years with relatively high losses, the costs under the reduced-sampling scenario will generally be lower than the minimum costs incurred when applying the base scenario.

Export consequences

This section makes a qualitative assessment of the export consequences based on the results of the epidemiological model. The section focuses on consequences for potato export. Figure 4 shows the distribution of the yearly number of exported infected but undetected seed lots, for both scenarios. Under the base scenario, there is a probability of less than 15 % that one or more infected seed lots are unintentionally exported, and the observed number of seed lots exported almost never exceeds five per year. For the reduced-sampling scenario, the probability of exporting one or more infected lots is increased to more than 60 %. Moreover, there is a considerable probability that the number of exported infected seed lots exceeds five. Thus, under the reduced-sampling strategy, the likelihood for an importing country to receive an infected lot from the Netherlands is relatively high compared to the base scenario.

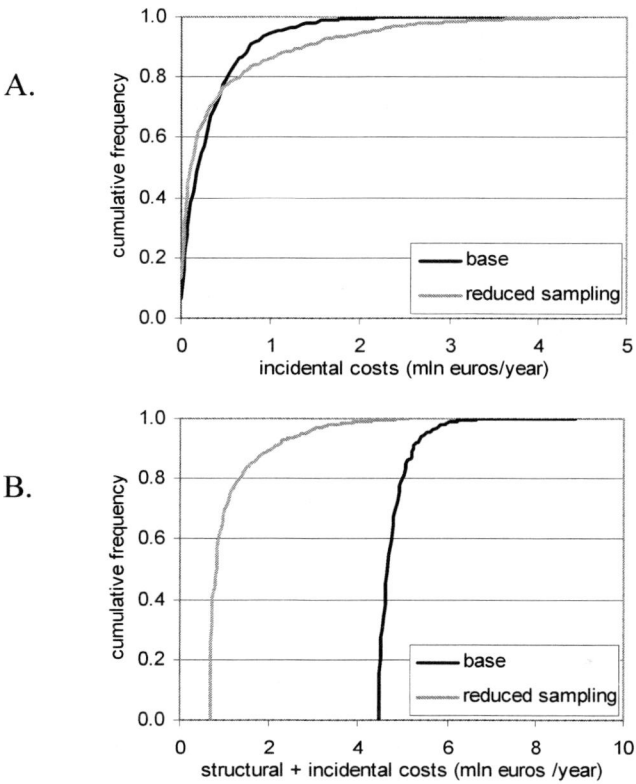

Figure *3. Cumulative frequency diagram of incidental costs (a) and total short-term (=structural + incidental) costs (b) per year, for the base scenario and the reduced-sampling scenario*

A large share of seed potatoes is exported to countries with a summer climate that is relatively conducive to brown rot (NAO 2004). It is likely that infected seed lots replanted in these countries will show visual symptoms of brown rot soon after they are replanted, resulting in detection. An incidental brown-rot detection in Dutch seed potatoes by another country will not immediately affect the export of Dutch seed potatoes. In contrast, at a level of infected exports as observed for the reduced-sampling strategy, the reliability of the disease-free status of Dutch seed potatoes may become at stake. Decreased confidence in the quality of Dutch seed potatoes will initially cause importing countries to require a more intensive testing policy in the Netherlands. Ultimately, it can result in a reduced export of this product. Since the elasticity of the domestic-demand curve for seed potatoes is rather low and the amount exported is about three times as high as the domestic demand, even a small reduction in export will have considerable consequences for the Dutch seed-potato price. As production of seed potatoes and ware and starch potatoes are strongly correlated, also the prices of ware and starch potato will be affected.

A decrease in potato prices will have economic consequences that are of a much higher magnitude than the short-term costs. For instance, lowering the seed-potato prices by 0.1 eurocent per kg already results in a loss over one million euros per year. Consequently, when taking into account the long-term consequences of the two scenarios described in this chapter, it is not so evident anymore that the reduced-sampling scenario is more cost-effective than the base scenario.

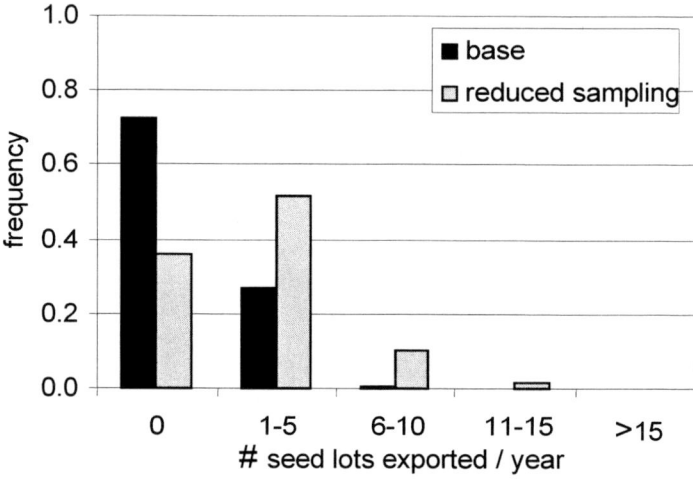

Figure 4. Distribution of the number of infected seed lots exported from the Netherlands per year, for the base scenario and the reduced-sampling scenario

OUTLOOK

In this chapter, we have introduced a new modelling concept for evaluating the cost-effectiveness of strategies for controlling brown rot in the plant production chain. Simulation results of brown-rot dynamics for the base scenario reflect the current knowledge of, and experience with brown rot in the Dutch potato production chain. As expected, reducing the sampling frequency of seed potatoes leads to an overall increase in brown-rot prevalence and, in particular, in the yearly variation in the number of brown-rot infections. The incidental costs of brown rot under reduced sampling are lower in most cases, but occasionally far exceed the incidental costs incurred under the base scenario. However, the yearly structural costs of the base scenario are more than six times higher than those of the reduced-sampling scenario, making the reduced-sampling scenario the more cost-efficient of the two scenarios in the short run. Yet, when including the export consequences in the model, this conclusion no longer holds. Although not quantified, it is very plausible that the reduced-sampling scenario leads to long-term losses as a consequence of reduced image of Dutch potatoes, which may more than cancel out its short-term economic benefits. In the near future, a method will be developed to analyse the export losses quantitatively.

An aspect that has not been discussed in this chapter is the issue of 'who has to pay'. The distribution of the costs over the different categories of stakeholders involved, as well as the distribution of costs between individuals of the same category, can affect the effectiveness of a strategy. If the majority of the costs are incurred at one level of the production chain, it is likely that these stakeholders will not support a specific control strategy. The same holds if within one level of the production chain some individuals have a much higher share in the costs than equivalent individuals. Lack of support for an imposed control strategy decreases the collaboration in the implementation of that strategy and encourages undesirable or illegal behaviour, which decreases the effectiveness of that strategy. Such behaviour can be avoided through surveillance, which increases the costs of the strategy. Another way to decrease the likelihood of illegality is to create an opportunity for stakeholders to share the costs or decrease the risk of incurring extremely high individual costs. For instance, in the Netherlands, potato growers can insure their crop against brown rot (not included in the economic analysis). Insured growers affected by brown rot receive a compensation for most of the incidental costs they incur. This compensation is paid from the insurance premiums of all growers; the incidental costs are thus shared among all insured growers.

The modelling concept presented here has a number of features, which altogether distinguish it from existing bio-economic models. Firstly, it focuses on an entire plant production chain. This has consequences for the modelling approach to describe disease dynamics, as these are not purely dominated by the biological behaviour of the pathogen anymore. The populations of plants that comprise the units in these chains behave as aggregated individuals for which the concepts of individual-based modelling apply. Secondly, the model is stochastic, and simulation results do not only show the average brown-rot prevalence and economic consequences but also their variations from year to year. The results show that this variation between years plays an important role when evaluating the cost-effectiveness of control strategies. Moreover, export consequences are shown to be important. Quarantine disease often cause minimal damage to host crops, in which case control and eradication measures have hardly any direct benefits. Yet, when taking into account what might happen when controlling the disease to a much lower extent, there is a significant benefit of 'avoiding even higher costs'. Finally, the model is spatially explicit, which increases the imaginative power of the model and allows for evaluating the effect of regional impositions of measures.

The model can be used, amongst others, to study the effect of measures concerning sampling strategy, farm management, or other factors that possibly affect brown-rot prevalence and dispersal. Thereby, it can be of great support to authorities and policy makers who are responsible for the implementation of effective and efficient control policies. With some adaptations, the model could be used to study control options for other 'chain-related' potato diseases, such as ring rot (*Clavibacter michiganensis* subsp. *sepedonicus*) and blackleg or soft rot (*Erwinia carotovora* subspecies). The conceptual framework presented in this paper is generally applicable to diseases in other production chains and other countries.

REFERENCES

Acock, B and Reddy, V.R., 1997. Designing an object-oriented structure for crop models. *Ecological Modelling*, 94 (1), 33-44.

Breukers, A., Hagenaars, T.omas, Van der Werf, W., et al., 2005. Modelling of brown rot prevalence in the Dutch potato production chain over time: from state variable to individual-based models. *Nonlinear Analysis: Real World Applications*, 6 (4), 797-815.

Breukers, A., Kettenis, D.L., Mourits, M., et al., in press. Individual-based models in the analysis of disease transmission in plant production chains: an application to potato brown rot. *Agricultural Systems*.

Dekkers, W.A., 2001. *Kwantitatieve informatie: akkerbouw en vollegrondsgroenteteelt 2002*. Praktijkonderzoek Plant & Omgeving, Lelystad.

Dijkhuizen, A.A. and Morris, R.S., 1997. *Animal health economics: principles and applications*. University of Sydney, Sydney.

Elphinstone, J.G., 2005. The current bacterial wilt situation: a global overview. *In:* Allen, C., Prior, P. and Hayward, A.C. eds. *Bacterial wilt disease and the Ralstonia Solanacearum species complex*. The American Phytopathological Society, St. Paul, 9-28.

Grimm, V., 1999. Ten years of individual-based modelling in ecology: what have we learned and what could we learn in the future? *Ecological Modelling*, 115 (2/3), 129-148.

Heesterbeek, J.A.P. and Zadoks, J. C., 1987. Modelling pandemics of quarantine pests and diseases: problems and perspectives. *Crop Protection*, 6 (4), 211-221.

Hendriks, H. and Höfte, M., 2004. *Rapportage evaluatie bruinrot/ringrot seizoen 2003: een beschrijving van de resultaten*. Plantenziektenkundige Dienst, Wageningen.

Huston, M., DeAngelis, D. and Post, W., 1988. New computer models unify ecological theory. *BioScience*, 38 (10), 682-691.

IPPC, 1999. *International Plant Protection Convention: New revised text approved by the FAO Conference at its 29th session, November 1997*. FAO, Rome. [https://www.ippc.int/servlet/BinaryDownloaderServlet/13742_1997_English.pdf?filename=/publications/13742.New_Revised_T ext_of_the_International_Plant_Protectio.pdf&refID=13742]

NAO, 2004. *Feiten en cijfers 2003*. Nederlandse Aardappel Organisatie, Den Haag.

Rushton, J., Thornton, P.K. and Otte, M.J., 1999. Methods of economic impact assessment. *Revue Scientifique et Technique de l'Office International des Epizooties*, 18 (2), 315-342.

Van Vaals, M. and Rijkse, H., 2001. *De Nederlandse Akkerbouwkolom: het geheel is meer dan de som der delen*. Rabobank International, Utrecht.

CHAPTER 6

RISK AND INDEMNIFICATION MODELS
OF INFECTIOUS PLANT DISEASES

The case of Asiatic citrus canker in Florida[1]

BARRY K. GOODWIN AND NICHOLAS E. PIGGOTT

*North Carolina State University, Box 8109, Raleigh, NC 27695, (919) 515-4620,
USA. E-mail: barry_goodwin@ncsu.edu*

Abstract. Asiatic citrus canker is an infectious disease that is a significant hazard to commercial citrus production in Florida. Our paper examines models of the risks of citrus canker transmission. The State of Florida currently has an active inspection program that checks every commercial grove several times each year. We use data from over 338,000 inspections over the 1998-2004 period. Simple models describing the risks of infection are used to evaluate risks and associated indemnity/insurance fund contribution rates. The risks are estimated for annual contracts which would pay producers a pre-specified indemnity in the event that their grove is found to be infected with canker.
Keywords: citrus canker; spatio-temporal risks; insurance models

INTRODUCTION

Florida had 748,555 acres of commercial groves in 2004 with the value of sales on-tree an estimated US$745.963 million (Florida Agricultural Statistics Service 2005). Florida is the largest citrus-growing state and accounts for 79 % of total U.S. citrus production. Figure 1 indicates that the estimated value of citrus production in Florida was $746 million in 2004, which represents a reduction from the most recent high of $1,108.523 million in 1999-2000 – a decline of 32.7 %. Total production in the 2003-04 crop year amounts to 291.8 million boxes with 242 million boxes of oranges (82.9 %), 40.9 million boxes of grapefruit (14.0 %), and 8.9 million boxes of other types of fruit (3.1 %) (Florida Agricultural Statistics Service 2005).

Citrus canker disease affects plants in varieties of citrus species and citrus relatives. The following citrus species have been identified as being 'highly susceptible': grapefruit, key/Mexican lime, Palastine sweet lime, and trifoliate citrus, sweet orange cultivars: Hamlin, Navel and Pineapple (Schubert et al. 2001). The disease is caused by a bacterial pathogen, *Xanthomonas axonopodis* pv. *citri*. Before the most recent detection in 1995, the disease was found in the U.S. on two

A.G.J.M. Oude Lansink (ed.), New Approaches to the Economics of Plant Health, 71–99.

previous occasions, in Florida and other Gulf Coast citrus-growing states in 1910 and on the Gulf Coast of Florida in 1986. Both of these previous infestations were reportedly resolved by eradication programs conducted by USDA and the affected states (USDA-APHIS 2005a).

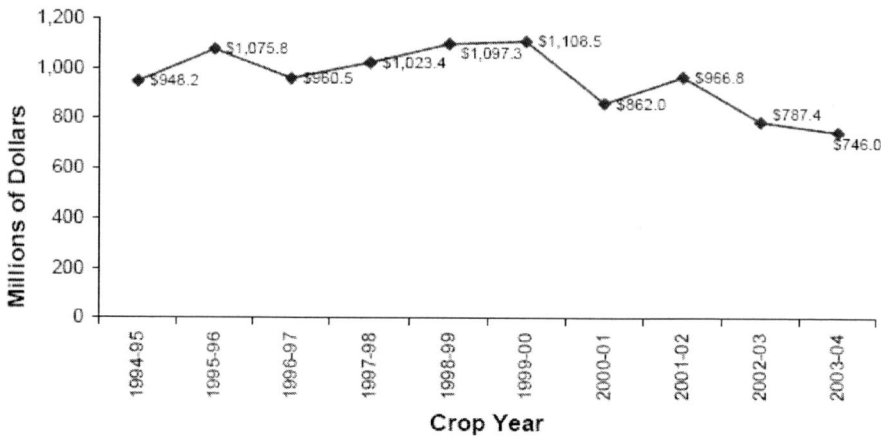

Figure 1. Florida citrus: value of sales on-tree, crop years 1994-1995 through 2003-2004

The current eradication program in Florida began in 1995 and has evolved into a program which involves separate infestations and different strains. It currently spans 13 Florida counties. In 1995 this current eradication program began to combat an Asiatic strain of citrus canker that was discovered in Florida in 1995 in a residential area near Miami International Airport[2]. Additional detections from this infestation culminated in an eradication program that included most of Miami-Dade County by 1998. Further, in May 1997 in what is believed to be a separate infestation, a different Asiatic citrus canker strain (thought to be connected to the 1986 infestation) was discovered in Manatee County in both residential citrus and commercial growing areas (USDA-APHIS 2005a).

Plants infected by citrus canker develop lesions on leaves, stems and fruit. These lesions ooze bacterial cells, making canker highly contagious. Canker can be spread rapidly by wind-driven rain, movement of equipment or workers that have come into contact with infected trees, or movement of infected or contaminated plants. These vectors of transmission, involving significant weather events and idiosyncratic movements of workers or people carrying contaminated plants, make containment a significant challenge. Once infection occurs it can take anywhere from 14 to 60 or more days for symptoms to appear. The bacteria can remain viable in lesions for several months (USDA-APHIS 2005a).

THE HISTORY OF CITRUS CANKER OUTBREAKS

Gottwald et al. (2001) point out that citrus canker has a long history dating back to the 1910s, when it entered from improved seedlings from Japan. Declared eradicated by 1993, a new infection was found in Mantee County, Florida in the late 1980s. This infection was thought to have been eradicated by 1994. Gottwald et al. (2001) explain that a new and separate outbreak occurred in urban Miami in 1995 and, at around the same time, a re-emergence occurred in the same area where the outbreak occurred in the 1980s. Gottwald et al. (2001) estimate that the 1995 Miami discovery near the airport spread from an initial 14-square-mile area to over 1,005 square miles in the metropolitan area plus an additional 260 square miles of urban and commercial citrus areas through the state. They point out that genomic analysis of bacterial isolates revealed that the majority of this outbreak was largely associated with the Miami discovery and therefore human-assisted movement must have been a factor in its transmission. Furthermore, in early 2000, a third distinct isolate of Asiatic citrus canker was identified in Palm Beach County. Therefore, at present there are at least three types of citrus canker that have been introduced in Florida in the most recent two decades (Gottwald et al. 2001). The U.S. Department of Agriculture (USDA-APHIS 2005b) provides a brief chronology of key events related to citrus canker over the period 1995 to 2003. This time-line consists of new discoveries of citrus canker over time, implementation of an eradication program, and legal challenges to this eradication program. In the discussion that follows, we highlight some of the key events as reported and identified by the USDA (USDA-APHIS 2005b).

In response to the September 1995 discovery of citrus canker in a residential area near Miami International Airport, the state of Florida and the USDA began administering surveys and implementing regulatory and control measures in the Miami-Dade County area. By June 1998, citrus canker had been found in Immokalee and in residential areas of Collier County. These infections were found to be related to the strain found earlier in Miami. Further, in the previous year, commercial groves in Manatee County were found to be infected and these infections were traced back to the strain that caused the 1986-94 infestations. In February 1999, an interim rule identified a federal quarantine area that had been expanded since the 1995 find to include 507 square miles of Broward and Miami-Dade counties, 68 square miles of Manatee county and 30 square miles of Collier county. A final rule that was published in July 1999 affirmed previous interims regulations that established a federal quarantine area encompassing Miami-Dade, Broward, Manatee and Collier Counties in Florida (USDA-APHIS 2005b).

Despite these quarantine efforts, the spread continued with additional discoveries of the Asiatic strain of citrus canker in residential areas of Hillsborough County in November 1999 and in lime groves in southern Dade County in January 2000. Schubert et al. (2001) reported that these discoveries led to destruction of almost half of the 4,000 acres of limes in the area due to exposure or infection. It was suspected that the disease was transferred via human activities from nearby residential areas to the north, with the oldest infections being detected in the highly susceptible pummelo fruit being grown in the vicinity of commercial lime groves. In

February 2000, the Florida Commissioner of Agriculture announced the implementation of a significant eradication program that would go into effect April 1, 2000. The key components of this program as described in USDA-APHIS (2005b) were as follows:

- decontamination of workers and equipment moving between groves;
- removal of all trees within a 1900-feet radius of an infected tree;
- establishment of a replacement program where residents whose trees that must be cut will be entitled to $100 voucher for the cost of a non-citrus tree; and
- establishment of a public-relations program.

In April 2000, several of the quarantine areas were also expanded (the Miami-Dade-Broward area and Collier County) and a new quarantine area of 106 square miles was established in Hendry County. At the same time, a sentinel survey program was initiated and there was a discovery of a third Asiatic strain of citrus canker on key limes in a Palm Beach residential area.

In October 2000, the Broward County Court cited improper rule-making and stopped the cutting of exposed trees within 1900 feet of infected trees. This was followed by an appropriation of $8 million in state funds in November to restore homeowners' property losses. These funds were in addition to the $100 vouchers already available for each tree lost. This also preceded proposed compensation to commercial growers for lost income due to the emergency control measures. In July 2001, a state administrative court found that the Florida Department of Agriculture exceeded its authority and therefore had to undergo an evaluation of its process of rule-making concerning the 1,900-feet cutting policy. Public hearings were held and in November 2001 a new rule extending the cutting of trees in proximity to exposed trees from 125 feet to 1,900 feet was implemented. These legislative efforts were challenged by Broward County, who filed briefs in administrative court during the same month countering the new rule. In March 2002, the state legislature passed a bill that was signed by the Governor of Florida, authorizing the removal of all citrus trees with the 1,900-feet area and permitting the use of blank search warrants. The Department of Agriculture and Consumer Services appealed the judgment in April 2002. In May 2002, a Broward County Circuit Court judge ruled that the eradication program that involved cutting exposed trees and using blank search warrants was unconstitutional since it violated constitutional search and seizure laws. At the same time, a Miami nursery won a restraining order to prevent the Department of Agriculture from removing calamondin trees. The significant amount of pending legal action led Florida Department of Agriculture officials to request permission to cut exposed trees in Palm Beach County in June 2002.

In July 2002, further litigious events transpired with the 4th District Court of Appeal ruling that attorneys could bypass the Court and go straight to the State Supreme Court due to the importance of the matter and its impact on the public. The Supreme Court in turn rejected this ruling and sent the action to the district court of appeals. Meanwhile in August 2002, citrus canker was discovered in Lee County, making fourteen counties that had positive finds since the 1995 discovery. The discovery was followed by the District Court of Appeals certifying a class action lawsuit by those who had been affected by the eradication program and who were seeking damages. By October 2002, new infections were found in Sarasota and Okeechobee Counties and a judge signed search warrants allowing mandatory inspections. In November and December of 2002, new quarantine areas were established in Orange and Lee Counties while areas in Collier and Hendry Counties were reduced in size. The first few months of 2003 saw more legal disputes which ultimately culminated with the Florida Supreme Court agreeing to hear an appeal from South Florida homeowners.

CITRUS CANKER PROGRAMS

Tree replacement payments

An interim rule was published on October 2000 providing eligible producers of commercial citrus payments to replace trees removed because of citrus canker (USDA-APHIS 2000). The payment was in the amount of $26 per tree, up to a maximum of between $2,704 and $4,004 per acre depending on the variety (Table 1). Per-acre payment caps were determined by the $26 per tree amount multiplied by the average number of trees per acre for a particular variety. This $26 payment per tree was determined by the USDA's Risk Management Agency (RMA) and took into consideration the costs of land preparation, replacement trees, labour for planting, and maintenance until the trees became productive (USDA-APHIS 2000). It was estimated that this program would compensate producers approximately $18.8 million with the payment of $26 per tree and an estimated 723,800 trees

Table 1. *Lost-production payment and tree replacement by variety*

Citrus varieties	Lost-production[a] payment[a] (a)	Maximum tree[b] replacement[b] (b)	Combined (a) + (b)
	Dollars per acre		
Limes	6,503	4,004	10,507
Orange, valencia and tangerine	6,446	3,198	9,644
Orange, navel*	6,384	3,068	9,452
Grapefruit	3,342	2,704	6,046
Other mixed citrus	3,342	2,704	6,046
Tangelos	1,989	2,964	4,953

*Source: USDA-APHIS (2002); USDA-APHIS (2000), includes early and midseason oranges
[a]Per-acre loss in the net present value; tree replacement cost has been deducted; per-acre income is determined by yield per tree (# boxes) multiplied by the price of a box less production costs per tree; the cash flow per tree is multiplied by the number of trees to determine per-acre net income
[b]Based on up to a $26 per-tree allowance; per acre caps were calculated by $26 times the varietal average number of trees per acre; the $26 per-tree allowance covers land preparation, replacement tree, labour for planting, and maintenance until the tree become productive

having been destroyed. However, the actual cost is estimated to be less because of the per-acre cap on payments.

Lost production payments

Tree replacement payments began in 2000 to compensate owners of commercial citrus groves who lost trees because of citrus canker. The lost-production payments went beyond the loss associated with the cost of the tree and compensated producers for the forgone income caused by the removal of commercial citrus trees to control canker. Owners of commercial citrus groves were made eligible if trees were removed because of a public order between 1986 and 1990 or on or after September 28, 2005 (USDA-APHIS 2002). Production payments are paid on a per-acre basis and vary across types of citrus trees, as is shown in Table 1. Limes have the largest payment at $6,503 per acre for lost production and a maximum payment of $4,004 per acre for tree replacement. Next are oranges, valencia oranges and tangerines with a payment of $6,446 per acre for lost production and a maximum payment of $3,198 per acre for tree replacement. Payments on navel oranges are slightly less with $6,384 per acre for lost production and a maximum of $3,068 per acre for tree replacement. Grapefruit and other mixed citrus fruits had considerably lower payment levels, with a lost production payment of $3,342 per acre and a maximum tree replacement payment of $2,704 per acre.

The rationale given for establishing production payments on a per-acre basis was that fruit output per acre is about the same, regardless of the number of trees. New groves have more, smaller and less productive trees, whereas older groves have fewer but larger and more productive trees. The per-acre amount is meant to reflect the approximate per-acre net income for each fruit variety, calculated by determining the revenue per tree and subtracting the production costs per tree to arrive at a net cash flow per tree, which is then multiplied by the number of trees per acre. USDA-APHIS (2002) explains that this per-acre value was calculated using a life-cycle approach with revenues and costs representing the productive life of a replanted grove. For limes this is 25 years. For other citrus varieties, the productive life was established at 36 years. The information utilized in these calculations employed data collected from the Florida Agricultural Statistics Service and the University of Floridas Institute for Food and Agricultural Sciences (UF-IFAS). If a producer purchased Asiatic citrus canker (ACC) crop insurance coverage and received an indemnity payment, lost production payments would be reduced by the amount of the indemnity payment. If the producer failed to purchase ACC if it was available, the per-acre production payment was reduced by 5 %.

Crop insurance

The Florida Fruit Tree Pilot Program began in 1996 and covered Dade, Highlands, Martin, Palm Beach and Polk Counties. Insurance was provided for the following tree types: orange, grapefruit, lemon, limes, all other citrus, avocados, carambolas and mangos. This policy is specifically aimed at tree stock rather than the fruit (another policy provides such coverage) and provides protection for damage to or destruction of trees. In 1998, a separate policy was developed for avocado and mango trees, which were dropped from the Florida Fruit Tree policy.

The policy initially insured against causes of loss that included excessive moisture and freeze or wind damage. An indemnity is triggered when damage to trees exceeds the chosen deductible. Coverage levels range from 50 to 75 % of the reference maximum price per tree. The insurance period ends the earlier of November 20 or upon determination of total destruction of insured trees (USDA-RMA 2005). In October 1999, the USDA-RMA announced that the Florida Fruit Tree Pilot Crop Insurance program for the 2000 crop year would be revised to allow producers to insure against losses to citrus trees arising from Asiatic Citrus Canker (ACC). The coverage area was expanded to 24 additional counties, making the pilot available to most commercial tree growers in an area that encompassed 29 counties. The ACC coverage was introduced as part of the standard policy but there are two sets of perils, standard and ACC, each determined separately. A producer in a county located without a quarantine zone qualifies for ACC coverage automatically. A producer in a county with a quarantine zone must obtain an ACC underwriting certification before coverage for ACC will be attached.

Table 2 documents that there was a significant increase in liabilities across the tree types and delivery methods (RBUP, CAT) in 1999-2005[3]. In 1999, total liabilities were only $156.8 million for all citrus in the Florida Fruit Tree policy. By 2005, this liability had increased to $1.141 billion. Initially in 1999, the most prevalent mode of delivery was through CAT coverage, which accounted for 91 % of total liabilities compared with the higher levels of coverage (RBUP), which only accounted for 9 %. The revisions in 2000 that included ACC as an insurable cause of loss transformed the preferred delivery. That is, a much larger proportion of trees were insured at higher levels of coverage than that provided by CAT, especially for the most susceptible citrus varieties – limes and grapefruits. The inclusion of ACC as an insurable cause of loss as well as the additional 24 counties that were included in 2000 explains the dramatic increase in liabilities, which rose from $156.8 million in 1999 to $697.3 million in 2000. By 2001, RBUP was the preferred delivery mode and this has remained the case with 63.4 % of liabilities being insured with RBUP in 2005.

Table 2 also documents another important characteristic of the current outbreak of citrus canker that is important to our empirical modelling work in later sections. Comparison of loss ratios across tree types suggests that some varieties are more susceptible and therefore more likely to be infected and receive an indemnity under this policy. Limes are the most notable, with loss ratios of 14.23 in 2000, 4.38 in 2001, 12.85 in 2002, and 6.63 in 2003 for the RBUP delivery[4]. These very large loss ratios as well as the rapidly declining total liability level for limes (which were $6.9 million in 2000 but only $83 thousand in 2005) reveals how adversely affected the lime groves have been by the current outbreak of citrus canker. The less susceptible oranges, which also happen to account for the largest share of total liability, have not had loss ratios for either delivery method that exceeded 1.0 in any insurance period

Table 2. *Florida fruit-tree crop insurance liabilities by type and mode of delivery 1999-2005*

Tree type	RBUP (a)	CAT (b)	Total (a)+(b)	%RBUP	%CAT	Loss ratio RBUP	Loss ratio CAT
		Dollars					
			1999				
All other	2,811,985	10,310,390	13,122,375	21.4%	78.6%	0.00	0.00
Carambola	23,071	328,662	351,733	6.6%	93.4%	0.00	0.00
Grapefruit	2,805,598	7,557,637	10,363,235	27.1%	72.9%	0.00	0.00
Lemon	0	0	0	0.0%	0.0%	0.00	0.00
Lime	458,456	2,577,002	3,035,458	15.1%	84.9%	0.00	0.00
Orange	7,962,313	121,946,556	129,908,869	6.1%	93.9%	0.00	0.00
Totals	14,061,423	142,720,247	156,781,670	9.0%	91.0%		0.00
			2000				
All other	15,443,152	28,301,459	43,744,611	35.3%	64.7%	0.00	0.01
Carambola	24,042	356,282	380,324	6.3%	93.7%	0.00	0.00
Grapefruit	56,248,255	45,846,180	102,094,435	55.1%	44.9%	0.38	0.79
Lemon	7,905	921,521	929,426	0.9%	99.1%	0.00	0.00
Lime	6,411,535	440,557	6,852,092	93.6%	6.4%	14.23	11.70
Orange	143,406,947	399,847,231	543,254,178	26.4%	73.6%	0.10	0.15
Totals	221,541,836	475,713,230	697,255,066	31.8%	68.2%		0.46
			2001				
All other	25,226,259	19,830,179	45,056,438	56.0%	44.0%	0.02	0.09
Carambola	67,320	174,723	242,043	27.8%	72.2%	2.06	0
Grapefruit	70,736,716	39,795,419	110,532,135	64.0%	36.0%	0.12	0.05
Lemon	1,689,194	0	1,689,194	100.0%	0.0%	0	0
Lime	4,072,664	63,959	4,136,623	98.5%	1.5%	4.38	0
Orange	319,596,759	349,139,103	668,735,862	47.8%	52.2%	0.21	0.14
Totals	421,388,912	409,003,383	830,392,295	50.7%	49.3%		0.19

Source: Federal Crop Insurance Corporation (http://www3.rma.usda.gov/apps/sob/)

Table 2 (cont.)

Table 2 (cont.)

Tree type	RBUP (a)	CAT (b)	Total (a)+(b)	%RBUP	%CAT	Loss ratio RBUP	Loss ratio CAT
		Dollars					
2002							
All other	35,503,321	20,725,293	56,228,614	63.1%	36.9%	0.00	0.00
Carambola	66,258	177,610	243,868	27.2%	72.8%	0.00	0.00
Grapefruit	88,630,388	41,334,491	129,964,879	68.2%	31.8%	0.00	0.07
Lemon	1,956,975	0	1,956,975	100.0%	0.0%	0.00	0.00
Lime	2,955,168	55,863	3,011,031	98.1%	1.9%	12.85	0.00
Orange	550,896,566	349,986,384	900,882,950	61.2%	38.8%	0.02	0.15
Totals	680,008,676	412,279,641	1,092,288,317	62.3%	37.7%		0.10
2003							
All other	32,902,961	19,106,230	52,009,191	63.3%	36.7%	0.10	0.03
Carambola	63,347	138,160	201,507	31.4%	68.6%	0.00	0.00
Grapefruit	81,166,014	35,757,250	116,923,264	69.4%	30.6%	0.26	0.07
Lemon	2,061,634	0	2,061,634	100.0%	0.0%	0.00	0.00
Lime	1,117,735	223,463	1,341,198	83.3%	16.7%	6.63	4.41
Orange	578,491,191	299,200,543	877,691,734	65.9%	34.1%	0.06	0.19
Totals	695,802,882	354,425,646	1,050,228,528	66.3%	33.7%		0.12
2004							
All other	30,100,685	19,560,289	49,660,974	60.6%	39.4%	0.49	0.09
Carambola	51,644	138,160	189,804	27.2%	72.8%	0.00	0.00
Grapefruit	77,462,930	40,678,332	118,141,262	65.6%	34.4%	0.55	0.01
Lemon	1,956,975	0	1,956,975	100.0%	0.0%	0.00	0.00
Lime	694,339	165,539	859,878	80.7%	19.3%	0.00	0.00
Orange	445,408,732	399,413,843	844,822,575	52.7%	47.3%	0.50	0.18
Totals	555,675,305	459,956,163	1,015,631,468	54.7%	45.3%		0.36
2005							
All other	37,987,207	17,763,543	55,750,750	68.1%	31.9%	1.21	0.20
Carambola	50,663	141,721	192,384	26.3%	73.7%	0.00	0.00
Grapefruit	92,406,857	33,973,728	126,380,585	73.1%	26.9%	2.21	2.37
Lemon	2,022,209	0	2,022,209	100.0%	0.0%	0.00	0.00
Lime	83,012	0	83,012	100.0%	0.0%	0.00	0.00
Orange	591,502,061	366,019,094	957,521,155	61.8%	38.2%	0.81	0.88
Totals	724,052,009	417,898,086	1,141,950,095	63.4%	36.6%		1.02

since 1999, with 2005 being the most adversely affected insurance period with loss ratios of 0.81 for RBUP and 0.88 for CAT. These liabilities and loss ratios highlight the importance of recognizing differences in the relative susceptibility across varieties as well as the spatial characteristics of the groves of different varieties when modelling the spatial and temporal risks of transmission.

BIOLOGICAL RESEARCH ON CITRUS CANKER

To model the spatial and temporal aspects of the risks of citrus canker transmission, it is critical to have a perspective on the biological research that has been conducted on citrus canker. In particular it is important to understand vectors of infection, the symptoms, rates of dispersion and other important characteristics that impact the spatial and temporal aspects of infection. In the discussion that follows, some of the key scientific research results on these topics are briefly discussed. A large number of these papers can be characterized as investigating a within-grove (or nursery) spread as opposed to spread across fields. The results of this research are useful in that they help to ascertain how the disease is spread. However, they are not directly applicable to our modelling effort in that we focus on the spread of the disease on a larger scale (such as across groves). The following brief discussion is by no means a complete review of the existing scientific knowledge on canker. Rather, it highlights some of the important findings that are pertinent to the empirical modelling in later sections of the chapter.

Graham et al. (2004) described the symptoms of citrus canker as distinct raised, necrotic lesions (localized death of living tissue) on the fruits, stems and leaves. The epidemiology involves bacteria spreading from lesions during wet weather and being dispersed at short range by splash, at medium-long range by windblown rain, and at all ranges by human assistance. The damage to the crop involves blemished fruits and defoliation. Importantly, Graham et al. (2004) point out that there are limited measures to prevent the spread of the bacteria[5]. Any blemished fruits are unmarketable and restricted from entering the market. This prohibition of market access is more significant than the actual losses pertaining to the yield of the crop.

Bock et al. (2005) used simulated, wind-driven rain splash to investigate the spread of the bacteria that causes citrus canker (*Xanthomonas axonopodis pv. citri*). The simulation involved electric blowers designed to generate turbulent wind and sprayer nozzles to produce water droplets entrained in the wind flow. Using this controlled environment, it was determined that citrus canker is readily dispersed in large quantities immediately after stimulus occurs. Furthermore, wind-driven splash was determined to have the capacity to disperse the inoculum for long periods and over a substantial distance.

Vernière et al. (2003) investigated environmental and epidemic variables associated with disease expression under natural conditions on Reunion Island. This research found that tissue age rating at the time of infection was a good predictor of disease resulting from spray inoculation on fruits and leaves and also on fruits following a wound inoculation. Mature green stems and leaves were also found to be highly susceptible after wounding while buds and leaf scars expressed the lowest susceptibility. Furthermore, temperature was also a significant factor in determining disease development.

Gottwald et al. (2002) investigated the spread of citrus canker in urban areas of Miami in the context of the effectiveness of the practice of removing exposed trees within 125 feet of infected trees in eliminating further bacterial spread. Several results from this work are of interest. It was established that a broad continuum of distance for bacterial spread was possible with maximum distances ranging from 12 to 3,474 meters in a period of 30 days. In addition, it was determined that the disease was best visualized 107 days following rainstorms with wind. Finally, this work showed that rapid spread of disease occurred across the regions studied in response to rainstorms with wind, followed by a filling in of disease on remaining non-infected susceptible trees through time by less intense rain storms.

Gottwald et al. (1992) compared spatial and spatio-temporal patterns of citrus canker infection in nurseries and groves in Argentina. This work involved innoculating the center plant in each plot with *Xanthomonas campestris pv. citri* and allowing the disease to progress for two growing seasons. Final disease incidence exceeded the 90-% level in all three nurseries and reached 69 % and 89 % for orange and grapefruit groves, respectively. Study of the proximity patterns reveals that some non-contiguous elements indicated the formation of secondary foci. Further these non-contiguous elements remained until the last few assessments, made every 21 days, before they eroded and the proximity patterns generally became larger and contiguous.

Spatial and temporal aspects of transmission

A key aspect of disease and pest contamination involves the spatial aspect of transmission. Pathways for transmission of diseases and pests generally have a spatial element. Thus, risks are highly correlated across space. In terms of modelling draws from distributions of yields in neighbouring geographic regions, it is clear that yield realizations from one region are certainly expected to be highly correlated with those in neighbouring areas. Spatial statistics play an important role in modelling the epidemiology of infectious diseases. An extensive literature, summarized by Alexander et al. (1988) and Rothenberg and Thacker (1992), has investigated spatial aspects of disease transmission. It is common in modelling spatial aspects of yield risk to assume that the correlation of risk declines with distance. This is certainly intuitive, though weather patterns are often directional and thus it is important that the directional aspects of spatial risk relationships be explicitly acknowledged when modelling the risks associated with invasive-species contamination.

Gottwald et al. (2001) outlined how the scientific basis for the eradication program now in place was initially based on data for Argentina, which indicated that canker could spread up to 105 feet with wind-driven rains. This led to an initial mandated removal and destruction of trees within a 125-foot radius; presumably the additional 20 feet was established as a precautionary measure. This 125-foot rule was ineffective and the disease continued to spread in urban areas and spread to several commercial citrus plantations in south Florida (Gottwald et al. (2001) citing

Gottwald et al. (1997)). This failure of the 125-ft. rule called into question the validity of this rule for three specific reasons that were spelled out by Gottwald et al. (2001) and reproduced here:

- the spread of citrus canker in a central Florida grove in the early 1990s was as much as 2,600 feet in a rainstorm;
- catastrophic weather (hurricanes and tornadoes) was documented by surveys to spread bacterium up to 7 miles; and
- the failure of the 125-ft. rule in citrus groves and urban areas to reduce the progress of the disease.

This failure and need for better information on the spatial characteristics of the spread led to collaboration between the Citrus Canker Eradication Program (CCEP) and the USDA-ARS and UF-IFAS to investigate and quantify the spatial patterns and dispersal of pathogens in a subtropical urban Miami setting. Gottwald et al. (2001) revealed that this epidemiological study took 18 months to complete and involved 19,000 healthy and diseased dooryard citrus trees in four areas: three in Dade County and one in Broward County, accounting for about 10 square miles. Figure 1 in Gottwald et al. (2001) illustrates the severity and contagiousness of this disease, showing how a single infected dooryard tree can lead to 1,751 infected trees over 18 months in a region of 12 square kilometres (3 kilometres north to south and 4 kilometres east to west).

This current outbreak of citrus canker presents an ideal case study for modelling risk since extensive data relating to transmission and the factors underlying risks have been collected. We shall utilize these data in an empirical model that identifies risks, potential losses, and appropriate premiums and contribution rates for an indemnification program. The State of Florida currently has an active inspection program that checks every commercial grove annually, with some groves being inspected several times each year. We use data from over 338,000 inspections over the 1998-2004 period. Simple models describing the risks of infection are used to evaluate risks and associated indemnity/insurance fund contribution rates. The risks are estimated for annual contracts which would pay producers a pre-specified indemnity in the event that their grove is found to be infected with canker. Implications for more sophisticated models of spatial/temporal risk relationships are also discussed.

RISK MODELS AND INSURANCE/INDEMNITY FUND CONTRACTS

As we have noted, a number of government programs have been directed toward providing compensation for those citrus producers affected by citrus canker. In the case of disaster relief, the assistance has been of an ad-hoc nature, with state and federal policy makers providing disaster payments in response to larger-scale infections. Current crop insurance programs have provided protection against tree losses resulting from canker infection. However, this protection has been part of an all-risk insurance plan. All-risk coverage may suffer from a number of shortcomings from the difficulties associated with measuring the risks from all possible hazards[6].

An alternative to all-risk insurance and ad-hoc indemnification plans is a specific-peril plan of protection. In this case, the task of quantifying risks is limited to a single peril. Protection is offered only for losses caused by this peril and thus actuarial considerations are limited to modelling only the risks associated with the particular peril being covered. Examples of specific peril policies include hail, flood and cancer insurance. It is often argued that such specific peril plans have an advantage in that it is easier to quantify the risks associated with a single hazard than to attempt to model the risks from all hazards, including those that may be unknown. Such an issue is especially pertinent to plant disease considerations, where the risks of new diseases that have not been previously experienced may be relevant.

The key element to any effective insurance or indemnification plan is comprehension of the risks associated with the hazards being covered. In insurance contracts, knowledge of this risk underlies the actuarially-fair insurance premium rate. The actuarially-fair rate corresponds to the rate (expressed as a percentage of total liability) that sets total premiums equal to total expected indemnities. For example, if I expect to pay $1,000 in a typical year on an insurance contract that covers up to $10,000 in total liability, the actuarially-fair premium rate will be 0.10 (or 10 % as it is more commonly expressed)[7]. In the case of an indemnification fund which could be funded by a levy on producers, the actuarially-fair premium rate is analogous to the checkoff rate (again expressed as a percentage of total liability) that must be charged in order to equilibrate expected payouts with contributions into the indemnification fund. The risk models needed to measure the actuarially-fair premium or checkoff rate usually are expressed in terms of the conditional probability density or cumulative distribution function underlying the outcomes being considered. For example, in the case of crop yield insurance, one is generally concerned with obtaining an estimate of the density describing crop yields. Consider an insurance plan that guarantees a certain proportion λ of expected yield μ. If yields y fall beneath the guarantee, losses will be compensated at a predetermined price of P. In this case, indemnities will be given by:

$$P \cdot max\{0, \lambda\mu - y\}. \tag{1}$$

It is convenient to express expected losses as a product of the probability of a loss and the expected level of y, conditional on y being below $\lambda\mu$. Without loss of generality, we can assume that all losses are paid at a price of one[8]. In this case,

$$E(Losses) = Pr(y < \lambda\mu)E(y \mid y < \lambda\mu), \tag{2}$$

where $E(\cdot)$ is the expectations operator and $Pr(\cdot)$ denotes the probability associated with the indicated event. If we denote the probability density function (pdf) of yields by $f(y)$, expected indemnity payouts will be given by:

$$E(Losses) = \int_0^{\lambda\mu} f(y)dy \left[\lambda\mu - \frac{\int_0^{\lambda\mu} yf(y)dy}{\int_0^{\lambda\mu} f(y)dy} \right], \tag{3}$$

where $\int_0^{\lambda\mu} f(y)dy$ is equivalent to the probability distribution function evaluated at $\lambda\mu$, which we denote as $F(\lambda\mu)$. The premium rate will be given by the ratio of $E(Losses)$ to total liability $\lambda\mu$:

$$Rate = \frac{E(Losses)}{\lambda\mu}. \tag{4}$$

In many insurance programs, loss occurs as an all-or-nothing event. For example, life insurance policies will pay a fixed amount only in the event of death, with no other provisions that could generate partial payments. Such a bond program simplifies the construction of insurance premium rates since the payout is predefined. In such a case, the expected loss is given by the product of the probability of a loss and the fixed payment made in the event of a loss. Likewise, the premium rate is equal to the probability of a loss occurring. Such a contract is suitable for situations such as the citrus canker case, where any exposure corresponds to a complete loss.

A number of important issues underlie such risk-modelling problems. A number of important questions pertain to the density function $f(y)$. A specific choice of the density function must be made. Goodwin and Ker (2002) discuss specification issues related to the distributional assumptions that must be made in modelling insurance contract parameters. As they note, one may choose to employ nonparametric density estimation techniques in cases where prior information about the parametric family governing the data-generating process is absent. Alternatively, a wide variety of parametric distributions are commonly applied to model parameters of insurance contracts. For example, crop yields commonly exhibit negative skewness, reflecting the natural biological constraints that govern maximum crop yields. Thus, a common choice for modelling crop yields is the beta distribution, which is capable of representing the negative skewness often observed for crop yields.

Recognition of the factors that loss events should be conditioned on is also an important component of risk models. For example, crop yields have exhibited significant trends over time and such trends must be explicitly recognized when assessing the risk of crops using data collected over time. Different crop practices are also an important determinant of risk. Irrigated crops typically have much lower yield risk than dryland production and thus any assessment of risk must be conditioned upon the crop production practice. To the extent that observable, deterministic factors are pertinent to risk, more accurate premium rates can be constructed by taking these factors into consideration. In the case of contracts to insure citrus canker risks, we know that factors such as fruit type and characteristics of the grove are important determinants of the risk of infection, and thus models of risk should be conditioned on such factors in order to produce accurate assessments of risk.

There are a number of operational considerations that must be considered when contemplating an insurance or indemnification program. One important factor involves the insurance period. A common insurance period is the calendar or crop year, where the terms of a contract are set prior to the beginning of the year and protection begins and ends with the beginning and ending of the year. In our analysis, we assume an insurance period corresponding to a calendar year. The period of insurance is important to how one models risk, since risks can only be conditioned on information available prior to the beginning of the insurance period. For example, it is widely recognized that hurricanes are an important causal factor related to citrus canker infection. However, in that it is impossible (or at least very difficult) to predict the occurrence of a hurricane at any single location in the following year, knowledge that prior infections were correlated with hurricane strikes is of little use in constructing insurance contracts. In contrast, we know that different fruit types have varying levels of infection risk. The type of fruit to be covered in year $t + 1$ is known at time t and thus the parameters of an insurance contract can be conditioned on fruit type.

An insurance contract must also specify the unit of insurance. Because of the diversification that comes with increasing size, risks are often lower as more aggregate units of insurance are defined. However, in cases such as citrus canker, where any exposure corresponds to a total loss, it is important that the unit be defined at a level consistent with the extent of loss upon exposure. Our data on canker inspections are given in terms of 'multiblock' units, which roughly correspond to individual commercial citrus groves. Multiblock units in our data average 14.7 acres in size and range from 0.05 to 510 acres.

In measuring risk and specifying insurance contract parameters, one must also decide upon the level at which risks will be measured. Alternative levels of aggregation may vary in terms of the stability of the premium rates implied as well as the accuracy of individual rates. In light of the spatio-temporal aspects of infection risks, the relative rarity of canker infections and the large number of multiblock observations, we utilize a degree of aggregation in our risk models. We considered two possible levels of aggregation. A common geographic designation based upon political boundaries is the 'Township–Range–Section' (TRS) definition. Townships are defined by township lines that run east and west every six miles, starting from a principal meridian and range lines that occur every six miles north and south of a principal meridian. Each 36-square-mile township is then divided into 36 individual square-mile sections. These designations were often determined many years ago as land was initially surveyed and thus may be subject to a number of errors or may reflect other difficulties associated with the initial surveys.

The dispersion of multiblock units used in our analysis and the TRS boundary lines of Florida is presented in Figure 2. Multiblock units, representing commercial citrus groves, are identified by the small shaded areas. The TRS boundaries are also identified. A limitation associated with using the TRS boundaries to identify insurable units is immediately obvious – some of the multiblock units are located outside of townships. This occurs in South Florida. The irregularity in the size and shape of TRS units may also make their use for defining units of homogeneous risk questionable.

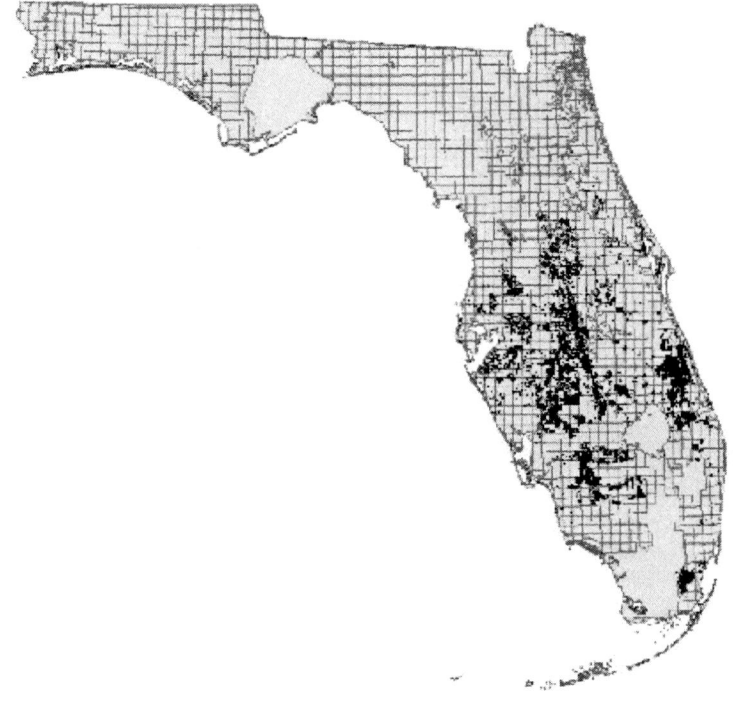

Figure 2. *Multiblocks and TRS designations*

In light of the limitations associated with the TRS units, we chose to identify our own insurable units based on an evenly spaced grid that covers the entire commercial citrus-growing region of Florida. We chose a grid defined by 10-km^2 units. The resulting grid is presented in Figure 3. As is true of the TRS designations, the groupings are ad hoc and other possible group definitions could have advantages. However, this approach was compared to grids of alternative sizes and found to perform well in the analysis that follows and to produce robust results.

Finally, our approach requires that we adequately incorporate any measurable factors that can be used to condition the risk of infection. Recall that only those factors that can be measured prior to the beginning of the insurance period are useful in conditioning the risk of infection. An important aspect of citrus canker, as with any infection disease, is that infection is spread through exposure to the infectious agent. We know that infection risk is subject to important spatial and temporal correlation factors. In particular, proximity in a spatial or temporal sense to existing infections raises the likelihood that a grove will be infected. We capture this relationship by considering the infections recorded in the previous year in all units having centroids that lie within 30 km of the centroid of the unit in question[9].

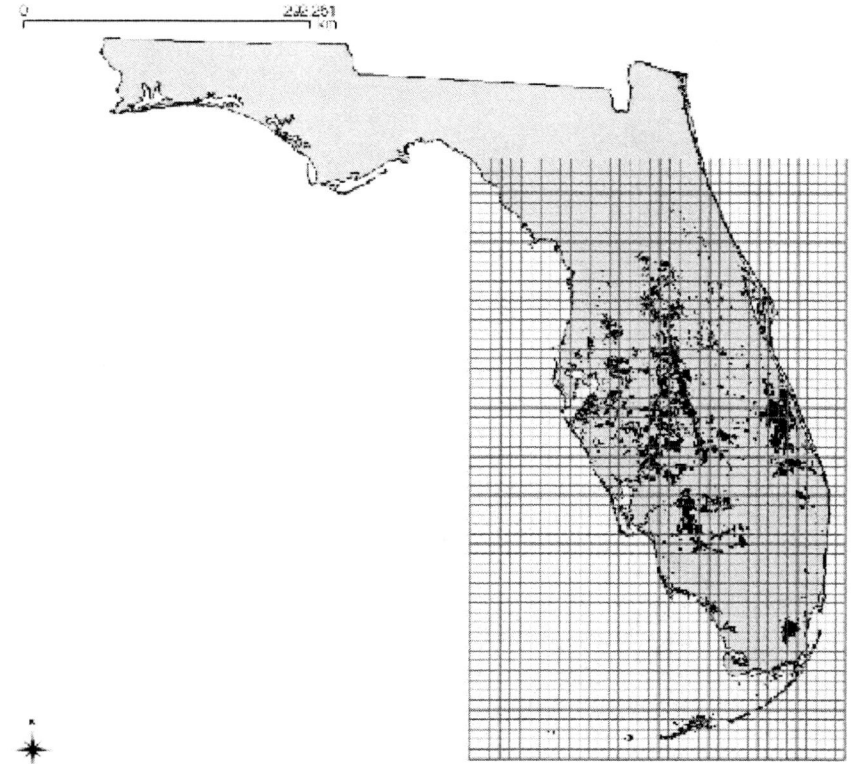

Figure 3. *Multiblocks and 10-km² unit grid*

Under these assumptions, we can view our risk-modelling approach to involve attempts to measure the conditional probability associated with citrus canker infection. This conditional probability can be expressed as:

$$Pr(y_{it}) = f(y_{it} \mid y_{jt-1}, \dots, y_{kt-1}, Z_{it}) + \in_{it}, \tag{5}$$

where $Pr(y_{it})$ corresponds to the probability associated with the event y_{it} (representing one or more canker infections in unit i in year t), y_{jt-1} is the infection status of neighbouring unit j in year $t - 1$, Z_{it} represents other predetermined factors conceptually relevant to the likelihood of canker infection, and $\in_{it}$ is a random residual error.

In order to make the transition to an empirical analysis, we must choose specific empirical models of the likelihood of infection. Our data are described in detail in the next section. Our measure of infection is the status of a particular multiblock unit at the time of its inspection – a discrete 0/1 indicator. In that we are applying the models to our aggregated 10-km^2 units, our measure of infection for the aggregate unit is the simple count of infections within the unit. Thus, we adopt two separate approaches to modelling the risk of infection. In the first, we consider probit models of the probability that one or more infections exist within a unit over a calendar-year period. Thus, we model:

$$d_{it} = f(X_{it}\beta),$$ (6)

using a probit model, where $d_{it} = 1$ if $y_{it} > 0$ and is zero otherwise. A second empirical approach makes use of the count nature of the infections data. We assume that the counts follow a Poisson process and model the count of infections within a 10-km^2 unit directly. The Poisson count model is given by:

$$Pr(y = Y) = \frac{e^{-\lambda}\lambda^Y}{Y!}, \text{ for } y = 0,1,2,....,$$ (7)

where λ represents the mean and variance of the random variable. We relate λ to explanatory variables through a logarithmic link function. Maximum-likelihood estimation procedures are used for both the probit and Poisson models.

DATA AND EMPIRICAL RESULTS

Our empirical analysis is based upon inspections data collected under the Florida Citrus Canker Eradication Program. The inspections data span 1996 through 2004. Data describing characteristics of the multiblock units and inspections reports were obtained from the Florida Department of Agriculture and Consumer Services Division of Plant Industry. The survey data report on the results of periodic inspections, which are made an average of 1.3 times per year on each multiblock. The data consist of reports on 338,226 inspections.

Discussion of data

Our unit of observation for our empirical analysis is the 10-km^2 unit of aggregation. The existing scientific evidence suggests that a number of observable factors may be relevant to the likelihood of infection. In particular, we know that certain fruit varieties are more susceptible to canker infection than others. Limes, lemons and grapefruits tend to be more susceptible than oranges and tangerines. We consider four variables representing the proportions of the citrus grove acreage in each aggregate unit devoted to particular fruit types – oranges, tangerines, grapefruit and all other fruits (which consist of limes, lemons, carambolas and other minor fruit varieties). It is also the case that there is considerable heterogeneity across our 10-km^2 units in the amount of citrus acreage. It is certainly the case that areas with more acreage are more likely to be found with infections. This occurs for two reasons. First, the infectious nature of citrus canker suggests that a denser concentration of citrus trees will correspond to a higher risk of infection. Second, there are likely to be more inspections in areas with more trees and thus a greater likelihood exists that canker will be found[10]. We include the total acreage of citrus surveyed in each unit as a conditioning variable in the probit and Poisson models. It is also the case that groves frequently have dormant acreage. Such dormant acreage could serve as a buffer against infection, at least to the extent that it insulates the fruit-bearing trees from the boundaries of the multiblock units. We include the proportion of total acreage that is dormant. Finally, we utilize a count of the total number of positive multiblock units in neighbouring units in the previous calendar year. Recall that neighbouring units are defined as any unit whose centroid is within 30 km of the unit of interest.

We utilize two indicators of a positive infection status. The first is simply an indicator of a positive finding in an inspection. The second indicator of infection is defined by a positive finding or any inspection in the two-year period following a positive finding. Current regulations under the Canker Eradication Program require that any grove found to be infected with canker must have its trees destroyed and then must remain fallow for a two-year period. This requirement assumes that canker spores remain infectious for up to two years after the trees are removed. Thus, our second measure assumes that all groves remain infected over the two years that follow a positive canker finding. Our dependent variables are the sums of these positive indicators over a calendar-year period.

Empirical results

The overarching goal of our models is to provide measures of the risk of canker infection which could be applied in the construction of insurance or indemnification plans. Perhaps the most straightforward approach to measuring such risk is to examine the locations of current and past infections and use spatial smoothing techniques to extrapolate exposure frequencies to provide infection probability measures. Of course, such an approach ignores any of the conditioning information that, as we have discussed, may be relevant to the risk of infection. Figure 4 presents infection probabilities obtained from spatial smoothing of historical infections in the inspections data. We used simple krigging procedures to estimate the probability surface. The surface indicates a higher probability of infection in the Miami area and in a few other areas that have experienced canker infections.

Such an approach ignores any conditioning information outside of historical infection locations that may be useful in assessing risks. In particular, as we have outlined in previous sections, plant pathology research has established that infection risks tend to be dependent upon a number of factors, including the type of fruit and timing of infections in neighbouring groves. Thus, it is likely that risk models that use such conditioning information may be much more informative. We estimated probit models of the discrete infection status (d_{it} = 1 for one or more infections and

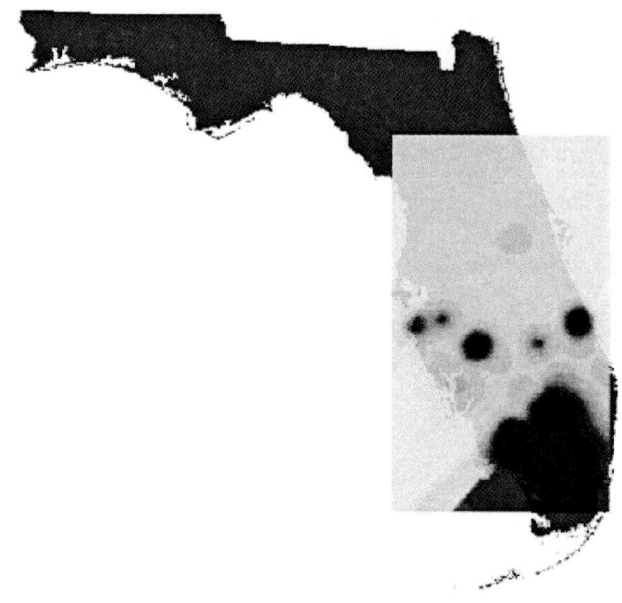

Figure 4*. Predicted probability surface using actual infection counts*

is zero otherwise). Recall that we utilize two measures of infection – a positive find and a positive status (the two-year period following a positive find). Table 3 presents summary statistics for measures of infection and other relevant explanatory factors. We present variable definitions and summary statistics both for the individual multiblock (grove) units and for the aggregate 10-km^2 block units in Table 3. There are 337,932 multiblock-level inspection observations and 2,380 annual aggregate block unit observations. Note that about 5.8 % of the aggregate observations have a positive infection status while only 2.5 % of the aggregate observations have positive finds. About 75 % of the citrus production is oranges, with other fruits accounting for smaller proportions.

Table 4 contains parameter estimates and summary statistics for the probit models of citrus canker infections. In both the positive-find and positive-status models, the parameters reveal a high degree of statistical significance, indicating the high degree of relevance of the conditioning variables. A likelihood ratio test of the joint significance of all of the explanatory factors is highly significant in each case. McFadden's LRI (also known as McFadden's R^2) ranges from about 0.178 to 0.200, again confirming the high degree of significance of the probit risk models. As expected, the risk models suggest that the likelihood of canker infection varies substantially across different fruit types. In particular, the parameter estimates suggest that oranges and tangerines have the lowest rates of infection, followed next by grapefruit and finally by other fruits (the default category), which consists of

Table 3. *Variable definitions and summary statistics*

Variable	Definition	Mean	Std. Dev.
Positive status	0/1 Indicator of a positive multiblock (up to 2 years after inspection)	0.0045	0.0667
Positive find	0/1 Indicator of positive canker survey	0.0006	0.0241
Acres	Size of multiblock unit (acres)	16.0347	23.9317
Orange acres	Orange acreage	13.0339	23.6261
Grapefruit acres	Grapefruit acreage	2.0410	8.4120
Tangerine acres	Tangerine acreage	0.5096	4.1285
Other acres	Other fruit acreage	0.0955	1.2898
Tangelo acres	Tangelo acreage	0.1918	2.1398
Lemon acres	Lemon acreage	0.0634	1.0315
Lime acres	Lime acreage	0.0986	1.4669
Dormant land	Dormant area (thousand square meters)	4.5286	26.5660
Land area	Total multiblock area (thousand square meters)	75.3921	107.3003
Unknown acres	Unknown acreage	0.0010	0.1081
Orange share	Orange acreage share	0.6885	0.4631
Grapefruit share	Grapefruit acreage share	0.1299	0.3362
Tangerine share	Tangerine acreage share	0.0395	0.1949
Other share	Other fruit acreage share	0.0144	0.1192
Tangelo share	Tangelo acreage share	0.0202	0.1406
Lemon share	Lemon acreage share	0.0103	0.1007
Lime share	Lime acreage share	0.0108	0.1034
Unknown share	Unknown acreage share	0.0001	0.0109
Dormant share	Dormant acreage share	0.0875	0.2825
....10-km^2 unit aggregates....................................			
Positive status	0/1 Indicator of a positive multiblock (up to 2 years after inspection)	0.0584	0.2346
Positive find	0/1 Indicator of positive canker survey	0.0252	0.1568
Orange share	Orange acreage share	0.7531	0.2922
Grapefruit share	Grapefruit acreage share	0.0917	0.1689
Tangerine share	Tangerine acreage share	0.0547	0.1292
Dormant share	Dormant acreage share	0.1098	0.2154
Positive neighbours (t-1)	Positive status units within 30km radius	1.9210	4.1317
Total acreage	Total unit acreage (hundred thousand acres)	0.0224	0.0412

^{a}Numbers of observations are 337,932 for multiblock units and 2,380 for 10km^2 units

lemons, limes and other minor citrus commodities. This finding is consistent with the implications of biological research, which has suggested that lemons, limes and grapefruits tend to be much more susceptible to citrus canker infections. It is important to point out that ignorance of fruit type in constructing and rating an insurance or indemnity plan would result in inaccurate rates, since important information relevant to the risks of infection would be ignored.

Table 4. *Probit model estimates of canker infection probabilities*[a]

Parameter	Estimate	Standard error	t-Ratio
...........................Model of positive status...			
Intercept	-0.7417	0.1364	-5.44*
Orange share	-1.4717	0.1524	-9.66*
Grapefruit share	-0.9383	0.2657	-3.53*
Tangerine share	-1.1121	0.3699	-3.01*
Dormant share	0.0280	0.1903	0.15
Positive neighbours (t-1)	0.0266	0.0095	2.80*
Total acreage	7.8289	0.7702	10.17*
Likelihood ratio test	180.77*		
McFadden's LRI	0.1706		
...........................Model of positive finds...			
Intercept	-1.0479	0.1647	-6.36*
Orange share	-1.5247	0.1938	-7.87*
Grapefruit share	-1.2725	0.3771	-3.37*
Tangerine share	-1.6060	0.7624	-2.11*
Dormant share	-0.3504	0.2632	-1.33
Positive neighbours (t-1)	0.0239	0.0126	1.90*
Total acreage	7.6514	0.9155	8.36*
Likelihood ratio test	111.95*		
McFadden's LRI	0.1999		

[a]Asterisks indicate statistical significance at the $\alpha = 0.10$ or smaller level

The probit models also suggest that the total amount of citrus acreage within each block is significantly related to the likelihood that inspections will reveal citrus canker. Again, this likely reflects the higher likelihood of infection in areas with a greater density of fruit trees as well as the greater likelihood that inspections will uncover one or more infections in areas with more trees. The proportion of grove area that is dormant has a negative, though not statistically significant relationship with infection risks.

Finally, the probit models confirm suspicions that infection risk tends to be spatially and temporally related to the realizations of other infections in neighbouring areas. The count of positive status multiblocks in all neighbouring units (defined by those units with centroids within 30 miles of the centre of the unit) has a positive and statistically significant effect on the probability of infection. This suggests that actuarially-fair premium or checkoff rates will be higher in areas in close proximity to infections in the preceding year.

Predictions from the probit models provide measures of the expected probabilities of canker infection. These probabilities are conditioned on fruit type, size, and the status of groves in neighbouring blocks in the previous year. Figure 5 presents a spatially smoothed (by krigging methods) representation of the predicted probability of canker infection. In comparison to Figure 4, which ignored all conditioning variables, a much richer picture of the risks of infection is offered by the probit models. In particular, the probit model predictions recognize the fact that infection risks are dependent upon the type of fruit, the density of production, and the status of neighbouring units.

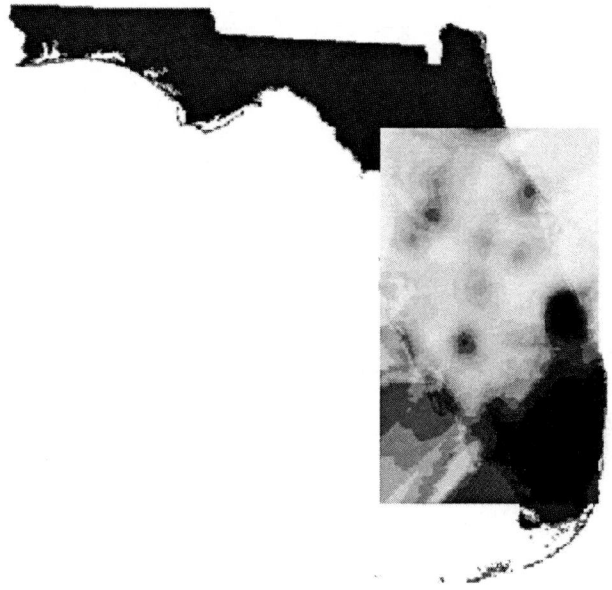

Figure 5. Predicted probability surface using probit model

The probit models provide statistically significant measures of the effects of various factors on canker infection probabilities. However, these models do not incorporate the degree of infection that may be present in the aggregate units. In particular, the probit estimates only account for the discrete status of canker infections and thus ignore the level or degree of infection. We know the number of positive inspections and multiblock units in each aggregate unit and thus a consideration of only the discrete status may ignore valuable information that could be used in modelling infection probabilities. To address this potential shortcoming, we also estimated Poisson count data regression models. The Poisson model parameter estimates and summary statistics are presented in Table 5.

Table 5. *Poisson logarithmic count model estimates of canker infection counts*[a]

Parameter	Estimate	Standard error	t-Ratio
............................Model of positive status...			
Intercept	1.6697	0.05	33.39*
Orange share	-3.6185	0.0741	-48.83*
Grapefruit share	-2.4731	0.1519	-16.28*
Tangerine share	-2.6196	0.2683	-9.76*
Dormant share	-1.0238	0.0953	-10.74*
Positive neighbours (t-1)	0.0435	0.0049	8.88*
Total acreage	12.1193	0.2244	54.01*
Pearson's χ^2	31,598.98*		
........................Model of positive finds..			
Intercept	-0.3445	0.1377	-2.50*
Orange share	-3.6594	0.2065	-17.72*
Grapefruit share	-2.6479	0.4420	-5.99*
Tangerine share	-3.0237	0.8595	-3.52*
Dormant share	-1.3306	0.2893	-4.60*
Positive neighbours (t-1)	0.0530	0.0132	4.02*
Total acreage	12.0353	0.6389	18.84*
Pearson's χ^2	7,927.27*		

[a]Asterisks indicate statistical significance at the $\alpha = 0.10$ or smaller level

The results are largely consistent with those obtained for the probit models. The estimates suggest that the risk of infection varies significantly across different fruit types, with oranges being the least susceptible, followed by tangerines, grapefruits and all other fruits. In contrast to the probit results, the share of acreage that is dormant now reflects a statistically significant negative relationship with infection risks. This is in accordance with expectations in that canker infection is expected to be less likely on dormant grove acreage. Dormant space may also serve to buffer existing fruit from future infections.

The Poisson models also confirm the probit results suggesting that infections in neighbouring units raise the likelihood that an infection will occur. Again, this reflects the infectious nature of citrus canker, which can be spread across space through a multitude of transmission means. Finally, the total scale of citrus acreage is again found to be significantly related to the likelihood of canker infection. This reflects the density factors and increased inspection frequency discussed above. One version of the Poisson regression model recognizes the fact that the counts may be measured over different possible numbers of positive events (i.e., in our case, different numbers of inspections). In such a case, adjustments may be made to recognize this different 'rate' of positive events. We do not pursue this estimation approach for two reasons. First, our inclusion of the total acreage as an explanatory factor explicitly accounts for differences in the rate of inspections, though in a more flexible manner than would be the case if an explicit adjustment were made to account for differing inspection rates. Second, we suspect that the density of citrus trees may have an important causal relationship with canker inspection risks and thus want to allow for a flexible relationship between the rate of inspections and the likelihood of canker infection[11]. Figure 6 presents the estimated probability of infection obtained from the Poisson model of positive infection status. Again, a much richer probability surface is implied by recognition of the conditioning variables.

Figure 6. Predicted probability surface using Poisson model

In all, the regression models confirm contentions that citrus canker infection risks tend to vary substantially across different fruit types, with risks the highest for lemons and limes and the lowest for oranges and tangerines. Density of production and infections in neighbouring areas also tend to be significantly related to infection risks.

Insurance/Checkoff premiums

The ultimate goal of our analysis is to use the estimated-risk models to construct measures of actuarially-fair premiums for an insurance or indemnity fund. In the context of our analysis, the actuarially-fair premium will be set equal to expected loss, which is given by

$$E\{Loss_{iJ}\} = F_i(\cdot) \cdot G_J(\cdot) \cdot \text{Payment, for } i \in J, \tag{8}$$

where i corresponds to multiblock i and J corresponds to aggregate 10-km^2 unit J. 'Payment' represents the payment to be made per acre in the event of a positive canker infection. In light of the calculations presented above, we assume that a unit of citrus stock is worth approximately $10,000 per acre and thus set the payment at this level[12]. The probit and Poisson models yield empirical measures of risk for the aggregate unit, given by $G_J(\cdot)$. We assume that all multiblock units within an aggregate unit having a positive status face an equal probability of infection and thus use the proportion of positive multiblocks in positive units as an empirical measure of $F_i(\cdot)$. This proportion is 0.0776.

Table 6 contains summary statistics for the estimated premiums for individual multiblocks. The premiums differ substantially across the alternative models, ranging from an average of $19.18 per acre for the probit model of positive finds to $229.63 for the Poisson model of positive canker status. The Poisson models may be suspect in light of the relatively rare nature of canker infections (less than 5 %). This may lead to a 'zero-inflation' problem that makes standard Poisson regression models suspect[13]. In all models, the highest premiums are in excess of $700 per acre each year – suggesting an infection probability of about 7 %. The dispersion of premiums is illustrated by Figures 5 and 6, which present the probabilities of infection determined from the aggregate-unit models. Note that the premiums are given by the product of the estimated infection probability at the aggregate-unit level, the multiblock conditional probability of infection (0.0776) and the payment ($10,000). The figures demonstrate that the premiums are highest in those areas that have realized the greatest incidence of canker infections. This includes those areas in the southern part of the state in the vicinity of Miami.

Table 6. *Summary statistics on premiums ($/acre) for canker coverage*

Model	Mean	Standard Deviation	Min	Max
Probit on Positives	44.53	62.67	10.43	746.22
Probit on Positive Finds	19.18	38.68	1.06	703.27
Poisson on Positives	229.63	176.68	38.75	776.00
Poisson on Positive Finds	48.26	83.40	3.74	775.80

CONCLUDING REMARKS

This analysis presents and evaluates models of the infection risks associated with Asiatic Citrus Canker in Florida citrus. We provide an overview of the history of citrus canker outbreaks in Florida. We also review biological aspects of citrus canker and discuss its relevance within the wider framework of invasive-species impacts on agriculture. We discuss methodological issues associated with the design of insurance and/or indemnification plan programs that would provide a form of 'self-help' risk protection for Florida citrus producers. The plan is presented in the form of a specific-peril program that offers to indemnify only those damages associated with citrus canker infections. The overarching goal of our analysis is to construct empirical risk models that allow us to quantify the risks of canker infection and uses these measures to identify actuarially-fair premiums or checkoff charges that should be paid for this protection.

We estimate probit and Poisson regression models that relate the risk of canker infection to a number of conditioning variables. Our models reveal that the risk of infection varies substantially across different types of fruit. The risk is lowest for oranges, followed then by tangerines and grapefruits. Minor citrus commodities, including limes and lemons, are found to face the highest risk of infection with canker. Our empirical models also reveal important spatio-temporal aspects of infection. Canker infection in neighbouring regions significantly raises the likelihood of infection. The size and density of citrus production in an area is positively related to the likelihood that canker infections will be found. The probit and Poisson model estimates are used to rate insurance/indemnity fund plans. The models suggest that the risks and thus premiums for protection are highest in the southern regions of Florida. This area is notable in that it has realized the highest incidence of canker infection.

A number of extensions to this research are currently being investigated. A wider array of empirical models that may be more flexible and more appropriate to the canker infection problem are currently being investigated. A specific interest is the suspected 'zero-inflation' problems associated with the relatively rare occurrence of canker in our data. In addition, several hurricane events that occurred in 2004 are very likely to be relevant to infections in 2005. Our analysis did not include data for the 2005 calendar year as our analysis was undertaken in mid-2005. As additional data are made available, we will focus modelling efforts on capturing the effects of the 2004 hurricanes, which are believed to have dispersed canker spores and thus led to a substantial increase in infections in 2005.

NOTES

[1] Goodwin is William Neal Reynolds Professor in the Departments of Agricultural and Resource Economics and Economics at North Carolina State University. Piggott is an associate professor in the Department of Agricultural and Resource Economics at North Carolina State University. This research was supported by the North Carolina Agricultural Research Service and by a grant under the PRESIM program of the Economic Research Service of the U.S. Department of Agriculture. We are grateful to Debra Martinez and Glen Gardner of the Division of Plant Industry in the Florida Department of Agriculture and Consumer Services for assistance with the data and our analysis. Direct correspondence to Goodwin at Box 8109, North Carolina State University, Raleigh, NC 27695, USA, e-mail: barry.goodwin@ncsu.edu.

[2] The current citrus canker infestation was detected in Florida on September 28, 1995. However, officials have identified five commercial citrus groves in Manatee and Highlands counties that were destroyed in previous limited outbreaks that occurred between 1986 and 1990 (USDA-APHIS 2002).

[3] RBUP indicates protection purchased at higher levels of coverage (above 50 % of yield and 60 % of price). CAT refers to catastrophic insurance coverage, which is provided to producers at a highly subsidized rate (consisting of only a small administrative fee).

[4] Loss ratios represent dollars paid out in indemnities per dollar paid in premiums.

[5] Gottwald and Timmer (1995) did find that use of windbreaks and copper bactericide can significantly reduce the temporal disease increase and spatial spread of citrus canker over time, with the windbreak being most effective.

[6] See Goodwin and Smith (1995) for a detailed discussion of contract design issues associated with all-risk crop insurance plans.

[7] Note that liability corresponds to payouts in a worst-case scenario. In other words, liability is defined by the limit on maximum indemnities. Premiums are typically expressed as the rate given by a percentage of total liability.

[8] Note that insurance premium rates are transparent to the price that losses will be paid at, since liability and indemnities are scaled by the same price, such that the ratio is unaffected by price. In an operational setting, however, it is possible that risks could be endogenous to price due to moral hazard. If the price is too high, individuals may undertake actions to increase their likelihood of collecting indemnities. We assume that such endogenous risks do not occur and thus that moral hazard is not an issue.

[9] The geographic centroid is the 'centre of gravity' of a geographic shape. In geometric terms, the centroid is the point at which a two-dimensional, planar shape would balance. In our units, the centroids are the exact canters of the 10-km^2 units.

[10] This raises an interesting point about our modelling exercise. We are not actually modelling the risk of infection but rather the risk that infection will be found by inspections. Of course, canker may exist and not be observed but such an event would not trigger indemnities under an insurance program and thus would not be relevant to the likelihood of payouts.

[11] This rate adjustment, often called an 'offset' adjustment, is analogous to entering the rate variable as a covariate with its parameter constrained to be one. We pursue a more flexible specification.

[12] The basis for this value of $10,000 per acre was formulated from the value of lost production and tree replacement, as is shown in Table 1. Note that, as long as risk is not endogenous to the payment level, risk and the underlying premium rate are transparent to the assumed payment level. Of course, a payment rate that is set too high may provide incentives for individuals to undertake actions that could increase their likelihood of collecting indemnities – the case of moral hazard.

[13] Current research is focusing on more general count data regression models, including models that explicitly address the overinflation problem.

REFERENCES

Alexander, F.E., Cartwright, R.A. and McKinney, P.M., 1988. A comparison of recent statistical techniques of testing for spatial clustering: preliminary results. *In:* Elliott, P. ed. *Methodology of enquiries into disease clustering.* London School of Hygiene and Tropical Medicine, London, 23-33.

Bock, C.H., Parker, P.E. and Gottwald, T.R., 2005. Effect of simulated wind-driven rain on duration and distance of dispersal of *Xanthomonas axonopodis* pv. *citri* from canker-infected citrus trees. *Plant Disease,* 89 (1), 71-80.

Florida Agricultural Statistics Service, 2005. *Citrus 2003-04 summary.* Florida Agricultural Statistics Service, Orlando. [http://www.nass.usda.gov/fl/citrus/cspre/cit92304.pdf]

Goodwin, B.K. and Ker, A.P., 2002. Modeling price and yield risk. *In:* Just, R.E. and Pope, R.D. eds. *A comprehensive sssessment of the role of risk in US agriculture.* Kluwer Academic Publishers, Norwell, 289-323.

Goodwin, B.K. and Smith, V.H., 1995. *The economics of crop insurance and disaster aid.* AEI Press, Washington.

Gottwald, T.R., Graham, J.H. and Schubert, T.S., 1997. An epidemiological analysis of the spread of citrus canker in urban Miami, Florida, and synergistic interaction with the Asian citrus leafminer. *Fruits,* 52 (6), 383-390.

Gottwald, T.R., Hughes, G., Graham, J.H., et al., 2001. The citrus canker epidemic in Florida: the scientific basis of regulatory eradication policy for an invasive species. *Phytopathology,* 91 (1), 30-34.

Gottwald, T.R., Reynolds, K.M., Campbell, C.L., et al., 1992. Spatial and spatiotemporal autocorrelation analysis of citrus canker epidemics in citrus nurseries and groves in Argentina. *Phytopathology,* 82 (8), 843-851.

Gottwald, T.R., Sun, X., Riley, T., et al., 2002. Geo-referenced spatiotemporal analysis of the urban citrus canker epidemic in Florida. *Phytopathology,* 92 (4), 361-377.

Gottwald, T.R. and Timmer, L.W., 1995. The efficacy of windbreaks in reducing the spread of citrus canker caused by *Xanthomonas campestris* pv. *citri. Tropical Agriculture,* 72 (3), 194-201.

Graham, J.H., Gottwald, T.R., Cubero, J., et al., 2004. *Xanthomonas axonopodis* pv. *citri*: factors affecting successful eradication of citrus canker. *Molecular Plant Pathology*, 5 (1), 1-15.

Rothenberg, R.B. and Thacker, S. B., 1992. Guidelines for the investigation of clusters of adverse health events. *In:* Elliott, P., Cuzick, P., English, D., et al. eds. *Geographical and environmental epidemiology: methods for small area studies.* Oxford University Press, London.

Schubert, T.S., Rizvi, S.A., Sun, X., et al., 2001. Meeting the challenge of eradicating citrus canker in Florida-again. *Plant Disease*, 85 (4), 340-356.

USDA-APHIS, 2000. *Q's and A's about citrus canker tree replacement (Plant Protection and Quarantine Fact Sheet).* Animal and Plant Health Inspection Service. [http://www.aphis.usda.gov/ lpa/pubs/fsheet_faq_notice/faq_phccanktree.html]

USDA-APHIS, 2002. *Q's and A's about citrus canker lost production payments (Plant Protection and Quarantine Fact Sheet).* Animal and Plant Health Inspection Service. [http://www.aphis.usda.gov/ lpa/pubs/fsheet_faq_notice/faq_phccankrpay.html]

USDA-APHIS, 2005a. *Emergency and domestic programs: citrus canker; background.* Animal and Plant Health Inspection Service. [http://www.aphis.usda.gov/ppq/ep/citruscanker/background.html]

USDA-APHIS, 2005b. *Emergency and domestic programs: citrus canker; chronology.* Animal and Plant Health Inspection Service. [http://www.aphis.usda.gov/ppq/ep/citruscanker/chronology.html]

USDA-RMA, 2005. *Fruit tree (pilot) Florida (Commodity Insurance Fact Sheet).* Risk Management Agency. [http://www.rma.usda.gov/aboutrma/fields/ga_rso/2005cropfactsheets/flcitrusfruit.pdf]

Vernière, C.J., Gottwald, T.R. and Pruvost, O., 2003. Disease development and symptom expression of *Xanthomonas axonopodis* pv. *citri* in various citrus plant tissues. *Phytopathology*, 93 (7), 832-843.

CHAPTER 7

HURRICANES AND INVASIVE SPECIES

The economics and spatial dynamics of eradication policies

ALBERT K.A. ACQUAYE, JULIAN M. ALSTON, HYUNOK LEE
AND DANIEL A. SUMNER[1]

*Department of Agricultural and Resource Economics, One Shields Avenue,
University of California, Davis, CA 95616, USA.
E-mail: acquaye@primal.ucdavis.edu*

Abstract. Citrus canker was established in Florida in the 1990s. The disease causes losses of yield and closure of some export markets. The U.S. government introduced an eradication policy in which growers are required to remove infected trees and receive compensation payments for doing so. Recent hurricanes have spread the disease and re-established it. This chapter examines the economic impacts of citrus canker in oranges and the eradication policy in Florida, taking into account the relationship between costs and benefits of eradication and the spatial and dynamic aspects of infestation. We evaluate both the costs of the disease and the benefits from eradication. In this evaluation we consider the implications of a future hurricane, which spreads citrus canker, for the decision about whether to adopt a strategy of eradicating initially, or after the hurricane, or both. We find that producers as a group benefit from both the disease and the eradication program, but at the expense of taxpayers, consumers and the nation as a whole. Producers benefit at the expense of consumers because both the disease and the policy to eradicate it reduce supply and drive up the price and the gross value of production. Producers also benefit at the expense of taxpayers, who pay to compensate them for their losses for having to remove trees under the eradication policy.
Keywords: citrus canker; eradication policy; invasive species; welfare economics

INTRODUCTION

Exotic pests can have significant effects on agricultural yields, product quality and costs of production, and they may introduce substantial costs through the loss of markets even when the production effect is limited. Exotic pests engender net social costs, partly because it is not worthwhile for individuals to prevent or eradicate them, even though it might be worthwhile for the industry or society. If the economic effects of an invasive species are large enough, and they involve externalities – either because the pest can spread from one farm to another, or because of effects on market access –, public policy by local, state or federal governments may be justified.

A.G.J.M. Oude Lansink (ed.), New Approaches to the Economics of Plant Health, 101–116.
© 2007 *Springer.*

The optimal choice of policy – to prevent, control or eradicate – will depend on the characteristics of the exotic pest, its costs, how it spreads, how easy it is to detect and eradicate, and the extent of the infestation. Intuitively, an eradication policy is more likely to be justified when the infestation is isolated or easy to isolate and the risks of spread are high; if the infestation is already spread over a wide geographic region eradication may be uneconomic. The extent of infestation may change over time, changing the policy calculus. For instance, random weather events may lead to a further dispersion of the exotic pest over a wider geographic region, thus increasing the cost of eradication and changing the nature of the externalities.

This chapter examines the case of citrus canker in oranges and the eradication policy in Florida. Citrus canker is a bacterial disease of most citrus species (including oranges, lemons, limes and grapefruit), caused by *Xanthomonas campestris* (=*axonopodis*) pv. Citri (*Xcc*). Severe infections may result in defoliation, unsightly blemishes on fruits, premature fruit drop, and general tree decline. Outbreaks of citrus canker occurred in Florida in 1986 1995 and 1997. The 1997 outbreak persists and the disease is a significant threat in California and other states (Jetter et al. 2003). The disease causes some loss of yields, but the main concern has been the potential loss of markets.

The U.S. government has sought to eradicate citrus canker since it was detected in 1995. The government inspects fruit and trees and mandates the removal of trees infected with citrus canker and trees in the surrounding areas both from residential and urban areas and commercial groves. These efforts can be undermined by hurricanes or other factors that spread the disease after it has been reduced to tolerable levels or confined to particular regions. Hurricanes in Florida in 2004 contributed to a major spread of citrus canker. As a result of these hurricanes many hundreds of thousands of additional commercial citrus trees were destroyed and hundreds of million of additional dollars were spent in compensation for growers in the broadened eradication efforts. In August 2005, further spread may have resulted from hurricane Katrina.

In previous work (Acquaye et al. 2005) we used a simple, aggregative, comparative static model to explore the implications of import tariffs and crop insurance subsidies for the consequences of citrus canker and eradication policies in the Florida orange-juice industry. That preliminary analysis abstracted from the dynamics of supply response, which are especially important for perennial crops like oranges, as well as the inherently spatial-cum-dynamic aspects of the spread of the disease. In contrast, in this chapter we use a spatially disaggregated dynamic model of farmers' planting decisions and market equilibrium. We model the economic impacts of citrus canker and eradication policies, taking into account the relationship between costs and benefits of eradication and the spatial and dynamic aspects of infestation, paying particular attention to the role of random weather events. In particular, we evaluate the implications of a future hurricane, which spreads citrus canker, for the benefits from introducing an eradication program. This aspect of our work is timely given recent events in Florida, but the issues that arise extend beyond citrus and beyond Florida to other exotic pests that may be spread by random factors.

An interesting feature of our results is that producers as a group can benefit both from an outbreak of citrus canker and from policies to eradicate it. This happens because the supply-reducing effects of the disease and the eradication policy both drive up the price of oranges, and it takes some time for supply response to undermine this effect. This *de facto* supply control aspect means that the interests of producers as a group directly oppose those of society as a whole. That finding has implications for the design of the eradication policy.

A SIMULATION MODEL

This section describes the elements of our simulation model of the market for Florida oranges. We model production of oranges at the level of counties in Florida. Complications arise because we explicitly model the age distribution of the population of orange trees, which is an important element of both the dynamics of supply response to price and the time path of the consequences of both the disease and the eradication policies. The dynamics of the industry are long-term, such that it is necessary to conduct market simulations over many years to see the full effects of disease outbreaks and eradication policies.

Supply of Florida oranges

The annual production of oranges in Florida depends on age-specific, weather-dependent, yields and the age distribution of trees. This may be specified as:

$$O_t = \sum_{c=1}^{N}\sum_{i=0}^{M} Y_{c,i,t} A_{c,i,t}, \tag{1}$$

where O_t is total production, $Y_{c,i,t}$ is the per-acre yield and $A_{c,i,t}$ is the area of trees in county c aged i years in year t. Normal yields of mature bearing trees in year $b+n$ (YM_{b+n}) are defined by yields in the base year (YM_b), an exponential growth rate (y) and a random annual proportional shock (μ):

$$YM_{b+n} = (1+y)^n YM_b (1+\mu_{b+n}). \tag{2}$$

A set of fixed weights (γ_i) define the age-specific yields as a fraction of mature yields:

$$Y_{i,b+n} = \gamma_i YM_{b+n}, \tag{3}$$

which are adjusted by a county-specific proportional yield shock (cc) associated with citrus canker:

$$Y_{c,i,b+n} = Y_{i,b+n}(1+cc_{c,b+n}). \tag{4}$$

The age distribution of trees is determined by past plantings (*PL*), and tree removals (*R*, determined exogenously in our model) and new plantings:

$$A_{c,i,t} = \begin{cases} PL_{c,t-1} & \text{if } i = 1 \\ A_{c,i-1,t-1} - R_{c,i-1,t-1} & \text{if } i > 1 \end{cases}. \tag{5}$$

The normal removal rates are assumed to be 2.3 % per year for trees less than 25 years of age, and 5.6 % per year for trees aged 25 years and older, based on the average in Table 1 of Brown and Stover (2002). We assume that removal of acreage associated with citrus canker is distributed proportionately across all age classes.

New plantings are based on profit-maximizing behaviour with a rational expectations formulation borrowed from Gray et al. (2005) and extended. Specifically,

$$PL_{c,t} = a_{0,c} + a_{1,c} E_{c,t} NPV. \tag{6}$$

In this equation, $PL_{c,t}$ is the number of acres planted in county c in year t, $E_{c,t}NPV$ is the expectation formed in time t of the net present value of planting an acre of oranges in county c in year t. Expectations are formed based on projections of the population of bearing trees, yields and demand. We conduct iterative stochastic simulations in which we derive distributions of projected outcomes for quantities and prices and so on, take expected values, and impose a model closure condition, requiring that the series of planting decisions is based on the expected net present values implied by the model given these planting decisions. Finally, $a_{0,c}$ and $a_{1,c}$ are parameters. Specifically, $a_{0,c}$ is the number of acres planted if the expected net present value from an acre of plantings is zero, and $a_{1,c}$ is the change in plantings for a unit change in expected net present value of an acre of plantings. Values for these parameters were derived by estimating a linear model for each county, in which annual county-specific plantings are regressed against budgeted estimates of the present value of a 50-year stream of profit per acre, using data for the years 1978 to 2002.

Demand for Florida oranges

Florida oranges may be sold for fresh consumption on either the domestic (*D*) or export markets (*E*), or for processing into orange juice (*J*). That is:

$$O_t \equiv D_t + E_t + J_t. \tag{7}$$

We include explicit demand equations for domestic and export fresh sales:

$$D_{b+n} = (d_0 + d_1 w_{b+n})(1+d)^n (1 - V_{b+n}) \tag{8}$$

$$E_{b+n} = \left(e_0 + e_1 w_{b+n}\right)\left(1+e\right)^n \left(1 - Z_{b+n}\right). \tag{9}$$

In these equations, quantities demanded depend on the wholesale price for fresh oranges (w), underlying exponential growth in the demands (at rates d and e), and demand shifters (V and Z) that reflect policy changes and other factors such as response to an outbreak of citrus canker. The fresh market is presumed to command a fixed premium (m) over the processing market price (w^p):

$$w_t = w_t^p + m. \tag{10}$$

The demand for processing use of Florida oranges is derived from the demand for orange juice from Florida taking into account processing costs and the yield of juice from oranges. We assume a fixed yield of juice (k gallons per box) and a fixed per-unit processing cost (w_0 \$/box) such that:

$$FOJ_t = kJ_t \text{ and } P_t = \left(w_0 + w_t^p\right)/k, \tag{11}$$

where FOJ is production of orange juice (in gallons) from Florida oranges (in boxes) and P is the price of Florida orange juice (\$/gallon). The demand for orange juice produced by Florida is equal to the demand for orange juice for current consumption (C) plus the demand for net changes in stocks ($S_t - S_{t-1}$), minus the supply of net imports (I) and the supply from other U.S. states (OOJ, which we treat as exogenous and fixed):

$$FOJ_{b+n} \equiv C_{b+n} + \left(S_{b+n} - S_{b+n-1}\right) - I_{b+n} - OOJ. \tag{12}$$

We define equations to represent each of the endogenous elements as follows:

$$C_{b+n} = \left(c_0 + c_1 P_{b+n}\right)\left(1+c\right)^n \tag{13}$$

$$I_{b+n} = \left(i_0 + i_1\left(P_{b+n} - \tau\right)\right)\left(1+i\right)^n \tag{14}$$

$$S_{b+n} = \left(s_0 + s_1 P_{b+n}\right)\left(1+s\right)^n, \tag{15}$$

where τ is the per-unit tariff on imports (29.72 cents per gallon single strength equivalent, SSE), and c, i and s are exponential growth rates in the functions. (We allow for the tariff but we ignore crop insurance, which was modelled by Acquaye et al. (2005) as equivalent to an output subsidy of 5.15 cents per gallon SSE.)

Substituting (13) – (15) into (12) yields an equation for the demand for Florida orange juice as a function of the price of orange juice and the parameters of demand.

$$FOJ_{b+n} = f_0 + f_1 P_{b+n} - S_{b+n-1} - OOJ, \text{ where}$$
$$f_0 = c_0(1+c)^n + s_0(1+s)^n - (i_0 - \tau i_1)(1+i)^n, \text{ and} \qquad (16)$$
$$f_1 = c_1(1+c)^n + s_1(1+s)^n - i_1(1+i)^n.$$

Substituting (11) into (16) yields an equation for the demand for processing use of Florida oranges (J) as a function of the price of oranges used for processing (w^P):

$$J_{b+n} = (FOJ_{b+n}) / k = \left(f_0 - S_{b+n-1} - OOJ \right) / k + f_1(w_0 + w_{b+n}^P) / k^2. \quad (17)$$

Equation (7) solves the model by equating supply (from equation (1)) with total demand (the sum of equations (8), (9) and (17)) for Florida oranges.

Parameters of the model and baseline prices and quantities

Tables 1 and 2 show the baseline quantities used to parameterize the model. The corresponding baseline price of oranges for processing was \$3.89/box, the price of fresh oranges was \$5.91/box, and the price of orange juice was \$1.22/gallon (all prices in 2003 dollars).

Table 1. *Production and utilization of Florida oranges (1997 – 2002 average)*

	Florida production	Share of Florida total	Share of Florida in U.S. total
	(*million boxes*)	(*percentage*)	
Fresh			
Domestic	10.01	4.5	24.04
Net export	0.70	0.3	6.55
Processing	209.87	95.1	95.90
Total	220 59	100.0	81.33

Source: Authors' computations based on data obtained from Florida Agricultural Statistics Service

Table 2. *U.S. production and consumption of orange juice (1997 – 2002 average)*

	Production	Share
	(*million gallons*)	(*percentage*)
Net imports	143.47	9.5
Florida production	1,340.67	88.3
Other U.S. production	56.52	3.7
Change in stocks	-22.61	-1.5
Total	1,518.04	100.0

Source: Authors' computations based on data obtained from Florida Agricultural Statistics Service

These baseline quantities and prices were combined with elasticities of supply and demand and other parameters (in Table 3) to initiate the model in 2004. The values for the elasticities were assumed, based on a review of estimates in the relevant literature, combined with knowledge of the industry and the structure of its markets. The critical parameter is the elasticity of demand for orange juice in the United States, and a value of -0.5 is representative of relevant estimates in the literature (e.g., see US International Trade Commission 2005).

Table 3. *Elasticities of supply and demand and growth rates*

Elasticities	
Elasticity of domestic demand for fresh Florida oranges	-1.00
Elasticity of demand for Florida exports of fresh oranges	-4.00
Elasticity of demand for orange juice in the U.S.	-0.50
Elasticity of demand for orange juice stocks in the U.S.	-0.50
Elasticity of supply of orange juice imports	5.00
Annual growth rates in supply and demand	*(percent)*
U.S. demand for fresh Florida oranges (d)	1.25
Export demand for fresh Florida oranges (e)	-4.68
U.S. demand for orange juice (c)	1.90
U.S. demand for orange juice stocks (s)	1.99
Yield of Florida oranges (y)	1.56
Import supply of orange juice (i)	-11.00

Source: Annual growth rates are past average growth rates over the period 1991 through 2002, computed by the authors using data obtained from Florida Agricultural Statistics Service

SCENARIOS SIMULATED AND SIMULATION RESULTS

The model is initiated in the year 2004 and runs for 50 years. In reality the disease is spread over time from tree to tree within and among farms and between urban back yards and farms, but here and for now, to simplify the problem and focus on the essential issues we treat the disease as either present within a region (involving several counties) or not. The eradication policy entails uprooting of infected trees, for which growers are paid compensation, financed by the federal government (Jetter et al. 2003). 'Eradication' does not eliminate the disease from the affected region, but for simplicity eliminates its impact. A hurricane causes the disease to spread from the infected region to neighbouring regions and re-establishes the disease in the initially infected region.

Benefits from eradication in the absence of hurricanes

In the first scenario we simulate the time path of prices, quantities and economic surpluses in the absence of citrus canker. In the second scenario we simulate the same variables with an outbreak of citrus canker in the Central region of Florida in

2011 (i.e., the 7th year of the simulation). A minor outbreak causes an immediate 10 % reduction in yield for 1 % of the orange acreage in the Central region of Florida. A severe outbreak causes an immediate 10 % reduction in yield for 10 % of the orange acreage in that region. In both cases, the outbreak results in a 50 % reduction in export demand for fresh Florida oranges. These impacts are sustained permanently in the absence of an eradication program.

In the third scenario, we simulate the same variables given an eradication program introduced in the year of the outbreak 2011. For simplicity, under the eradication program demand for fresh Florida oranges and yields are unaffected by the outbreak. The eradication program entails immediate removal of 15 % of the orange acreage (distributed at random across age classes) in the region, with no replanting for a further two years. Farmers receive compensation of $9,646 per acre lost as a result of eradication.

Table 4 summarizes the results of these scenarios in terms of the net present values of welfare impacts over the 50-year period 2004 – 2053, in 2003 dollars, expressed as the equivalent annual value of a perpetuity. The first two columns of numbers show the annual effects on welfare of the minor and severe outbreaks *without* eradication (i.e., comparing the first two scenarios). A minor citrus canker outbreak (column 1) causes a producer loss of $1.5 million, and a small loss to U.S. taxpayers, reflecting a slight reduction in imports of orange juice and thus in import tariff revenue. These losses are partially offset by a benefit to U.S. consumers of $0.9 million (from the lower domestic processing and fresh prices resulting from the reduction of demand for fresh exports), such that the net national loss is $0.5 million.

Column (2) shows the effects of a severe outbreak with no eradication. The citrus canker outbreak causes a producer gain of $5.5 million. This initially surprising result reflects the fact that overall demand for oranges is inelastic so revenue rises with the price increase caused by lower yields. Furthermore, the reduction in supply (from yield losses) is greater than the reduction in demand (from the loss of some fresh export markets) in this instance, such that producer revenue increases. U.S. taxpayers also gain slightly, reflecting an increase in imports and import tariff revenue because the loss in production is greater than the fall in exports of oranges. The resulting U.S. consumer loss from higher prices more than offsets the gains to producers (and taxpayers), resulting in a net national loss of $2.7 million per year.

Column (3) shows the effects of the outbreak *with* eradication (i.e., comparing the first and third scenarios). Relative to no outbreak, the citrus canker outbreak with eradication entails a producer gain of about $163 million, which is more than offset by losses to U.S. consumers and taxpayers totalling $188 million, such that national loss is about $25 million. The consumer losses here are caused by higher prices that result from the reduction of supply. The supply reduction is caused by the combination of the outbreak and the eradication program, where eradication has the larger effect. The taxpayer losses reflect both a slight increase in tariff revenue and significant expenditure on compensation to producers for tree removals.

Table 4. *Welfare change from a citrus canker event and eradication policy*

Changes in	Benefits from a minor event with no eradication (1)	Benefits from a severe event		Benefits from eradication (4)
		No eradication (2)	Eradication (3)	
	(annual values in millions of 2003 dollars over 50 years)			
Producer surplus	-1.45	5.50	162.80	157.31
Consumer surplus	0.94	-8.21	-176.00	-167.80
Fresh oranges	0.05	-0.29	-6.69	-6.39
Processed oranges	0.89	-7.91	-169.32	-161.40
Taxpayer surplus	-0.01	0.03	-12.17	-12.20
Tariff	-0.01	0.03	1.99	1.96
Compensation	0.00	0.00	-14.16	-14.16
Total domestic surplus	-0.51	-2.68	-25.37	-22.69
Foreign surplus	-0.25	-0.25	-0.25	0.00
World surplus	-0.76	-2.93	-25.62	-22.69

The benefits from eradication, in column (4), are computed as the differences between the values in columns (2) and (3) (i.e., equivalent to comparing the second and third scenarios). The eradication policy yields a benefit to producers of $157 million, a loss to U.S. consumers of $168 million and a loss to U.S. taxpayers of $12 million, such that the net national loss is $23 million. The consumer losses are again caused by higher prices that in this case result from the restoration of foreign demand but mostly from the reduction of supply caused by the eradication program.

The eradication policy in this instance benefits producers at the expense of consumers of fresh and processed oranges and taxpayers, and the losses to other domestic interest groups exceed the benefits to U.S. producers such that net national (and, indeed, global) benefits from the production and consumption of Florida oranges are reduced by the implementation of the policy. Ironically producers are compensated even though, as a group, they are net beneficiaries from eradication because tree removal causes price and revenue to rise.

A much simpler, static model of short-run supply and demand provides some intuition about the main factors behind these results. In Figure 1, *S* is the short-run supply of oranges in Florida, which is perfectly inelastic. An outbreak of citrus canker causes yield losses, and as a result supply shifts from *S* to *S'*. Since demand is inelastic, the effect of this shift is to increase both price and industry revenue (comparing the equilibrium at point *a* with the equilibrium at point *b*). At the same time, however, demand shifts from *D* to *D'*, reflecting the loss of some export markets. This shift results in a reduction in price and industry revenue (comparing point *b* and point *c*), offsetting the effects of the yield-related supply shift. As the figure is drawn the demand shift more than offsets the price and revenue effects of the supply shift (comparing points *c* and *a*), though it need not do so in every case we model, such that the price and producer revenue are lower than in the absence of citrus canker.

An eradication policy results in a further shift in supply to the left, from S' to S'', and a restoration of demand at D, such that the final equilibrium is at point e, with a smaller quantity but a higher price and a larger industry revenue than in the absence of the citrus canker outbreak and the eradication policy (i.e., comparing points a and e). Given an inelastic demand, producer surplus is greater at point e than at point a. Consumer surplus is lower as a result of the higher price, and the net effect, combining the consumer loss and the aggregate producer gain, is a loss equal to area $Q''eaQ$. In addition, however, taxpayers pay compensation to producers as part of the eradication program, such that the aggregate producer benefit is even greater, at the expense of both taxpayers and consumers. These types of effects drive the results in Table 4, though those results reflect a much more complicated dynamic supply and demand structure than we have used in Figure 1; the same factors also drive the further results that follow, which also involve complications from hurricanes.

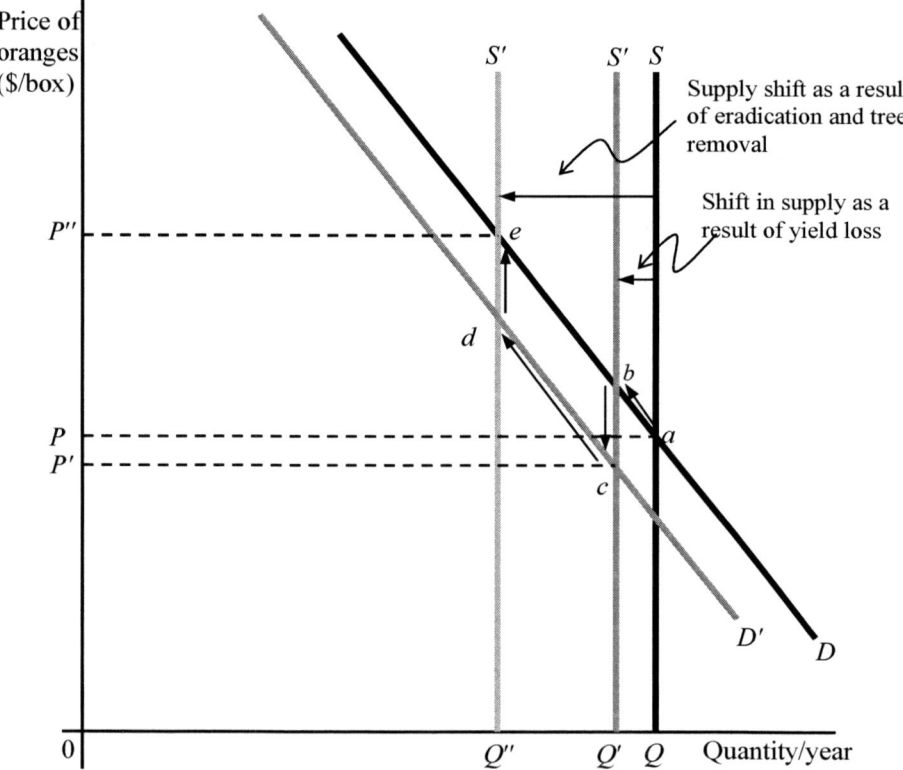

Figure 1. *Effects of a citrus canker outbreak and eradication policy in Florida*

Incorporating the effects of hurricanes

The cases in Table 4 do not allow for the effect of hurricanes. Further simulations replicate the second and third cases (i.e., a severe outbreak of citrus canker in 2011 with or without an eradication program) but with a hurricane in 2016. This timing of the hypothetical hurricane, five years after the outbreak, corresponds loosely to the actual history with an outbreak of the disease and introduction of an eradication program in the mid to late 1990s followed by a hurricane in 2004.

We assume the hurricane re-establishes citrus canker in the Central region – even if it had been "eradicated" some residual infection, sufficient to re-establish it, remains – and spreads the disease from the Central region either (a) to two other regions (the Northern and Western regions), if an eradication program *had* been introduced in 2011, or (b) to all six regions in Florida, if an eradication program *had not* been introduced in 2011. Given the scenarios of (a) an outbreak of citrus canker in 2011, after which the government may or may not introduce an eradication program, and (b) a hurricane in 2016, after which the government once more may or may not introduce an eradication program, we can contemplate four combined scenarios, each of which implies a different set of values for the parameters of the model. These scenarios and the corresponding parameters are summarized in Table 5. In every case, citrus canker causes a yield loss of 10 % on the infected acres, but the different scenarios have different numbers of acres affected and different implications for demand.

Column (1) in Table 5 shows the parameters that apply in the case when an eradication program was not introduced after the initial outbreak in 2011. After the hurricane, citrus canker affects 15 % of the orange acreage in every region of Florida, and, with no eradication program in 2016, this effect continues in perpetuity. Alternatively, if an eradication program is introduced in 2016, the rate of infection progressively declines from 15 %. At the same time, export demand progressively returns as embargoes are removed. The initial outbreak in 2011 results in a 50 % loss of export markets, and we assume that when an eradication program is introduced after the hurricane, the market gradually grows until 'normal' export demand is restored.

Column (2) shows the parameters that apply in the case when an eradication program is introduced after the initial outbreak in 2011. After the hurricane, citrus canker now affects only 10 % of the orange acreage in a smaller number of regions. If an eradication program is not introduced in 2016, this effect continues in perpetuity, but if an eradication program is introduced in 2016, the rate of infection progressively declines. Further, if an eradication program is not introduced the new infection results in a 50 % reduction in export demand for fresh Florida oranges.

Table 5. *Parameterization of the model under various scenarios*

	Without eradication in 2011 (1)	With eradication in 2011 (2)
Areas infected after hurricane in 2016	All six regions in Florida	Central Florida plus two other regions
Yield consequences for infected acreage	10 % yield loss on infected acreage	10 % yield loss on infected acreage
Consequences without eradication in 2016		
Proportion of acreage infected in affected regions	15 % of acreage in perpetuity	10 % of acreage suffers a 10 % yield loss in perpetuity
Effects on export demand for fresh oranges	50 % reduction in 2011 is sustained in perpetuity (i.e., no new demand shift caused by the hurricane)	permanent 50 % reduction in demand resulting from the new spread caused by the hurricane
Consequences with eradication in 2016		
Proportion of acreage infected in the affected regions	15 % of acreage initially, but declining progressively by half of the previous level each year to approximately zero by 2030	10 % of acreage initially, but declining progressively by half the previous level each year to approximately zero by 2030
Effects on export demand for fresh oranges	Export demand grows progressively to reverse the 50 % reduction in demand from the 2011 outbreak by 2021	No demand shift
Effects of eradication on orchards	20 % of acreage in affected counties removed from production, no replanting for two years and compensation of $9,646 per acre	15 % of acreage in affected counties removed from production, no replanting for two years and compensation of $9,646 per acre

As before, the eradication program itself also entails reductions in supply associated with mandated removal of trees. We assume that when an eradication program applies after the hurricane, 15 % of the orange acreage in the affected regions will be removed if an eradication program had been previously established, or 20 % of the orange acreage in the region if an eradication program had not been previously established. In each case, when trees are condemned and removed under the eradication program growers must not replant for two years and they receive compensation of $9,646 per acre lost.

Benefits from eradication in the presence of hurricanes

Based on these simulations, we evaluate the effects of an eradication program introduced after the hurricane in 2016, with and without an initial eradication program after the outbreak in 2011. The results from the various combinations of policies are summarized in Table 6.

Column (1) in Table 6 replicates the last column in Table 4, to show the benefits from eradication after the initial outbreak in 2011, if there will not be a subsequent hurricane. All of the other columns in Table 6 refer to simulations with a hurricane in 2016. The results in column (1) were explained above. The key point is that eradication is beneficial for producers but expensive for taxpayers, mainly because of compensation payments, and for consumers, because of reduced supply and higher prices, and ultimately for the nation as a whole. Reflecting the same main factors at work, the same pattern of results can be seen in all of the other columns except column (3), for reasons that will be explored next.

Columns (2) and (3) show the benefits from eradication following the outbreak in 2011 given that there will be a subsequent hurricane-induced outbreak in 2016 that will be either eradicated (column 3) or not (column 2). Comparing column (2) with column (1) a hurricane that reintroduces the pest in 2016 reduces the benefits from eradication in 2011. This makes sense because the reintroduction of the pest effectively eliminates the stream of benefits from the eradication in 2011 that would otherwise have continued through to the end of the simulation period 2053. Column (2) shows that if we are not going to eradicate citrus canker following a hurricane in 2016, an eradication program in 2011 will still involve a net cost to consumers, taxpayers and the nation as a whole even though it would benefit producers. These, and the other effects in column (2), are similar to those of an eradication program in 2011 in the absence of a hurricane in column (1), but muted.

In contrast, all of the entries dealing with domestic and world welfare in column (3) are of the opposite sign to their counterparts in columns (1) and (2). Column (3) shows that if policy-makers are confident that the nation will eradicate in 2016 following the hurricane, it would benefit consumers, taxpayers and the nation as a whole also to eradicate in 2011. Column (2) indicates that if we are not going to eradicate in 2016 it would not benefit consumers, taxpayers or the nation as a whole also to eradicate in 2011. Eradication in 2016 reverses the consequences of eradication in 2011, yielding large gains to U.S. consumers, with U.S. taxpayers being saved the burden of large eradication costs, and large costs to U.S. producers. The essential story here is that eradication in 2016 is costly to consumers, taxpayers and the nation (as shown in columns 4 and 5), albeit beneficial to producers. Eradication in 2011 reduces all of those positive and negative consequences from eradication in 2016, and for this reason the general impacts of eradication in 2011 given eradication in 2016 are opposite those of eradication in 2011 given no eradication in 2016.

A.K.A. ACQUAYE ET AL.

Table 6. Effects of a hurricane on the benefits from eradication

	Benefits from eradication in 2011 given			Benefits from eradication in 2016 given	
	No hurricane in 2016	Hurricane in 2016		Hurricane in 2016	
	No hurricane in 2016	No eradication in 2016	Eradication in 2016	No eradication in 2011	Eradication in 2011
	(1)	(2)	(3)	(4)	(5)
Changes in	*(annual values in millions of 2003 dollars over 50 years)*				
Producer surplus	157.31	139.27	-57.64	441.51	244.60
Consumer surplus	-167.80	-145.17	79.82	-533.78	-308.80
Fresh oranges	-6.39	-5.57	1.52	-16.84	-9.75
Processed oranges	-161.40	-139.59	78.29	-516.94	-299.05
Taxpayer surplus	-12.20	-11.90	16.35	-52.65	-24.39
Tariff	1.96	1.95	-0.23	5.32	3.14
Compensation	-14.16	-13.85	16.58	-57.96	-27.53
Total domestic surplus	-22.69	-17.80	38.53	-144.92	-88.59
Foreign surplus	0.00	-0.06	-0.03	-0.08	-0.05
World surplus	-22.69	-17.86	38.50	-145.00	-88.64

The last two columns show the benefits from eradication after the hurricane-induced outbreak in 2016. The magnitude of the welfare consequences of eradication in 2016 are affected by the decision to eradicate or not following the initial outbreak in 2011. This is so because the disease is much worse in 2016 if there was no initial eradication (it extends to a much broader area, the rate of infection is worse, and the rate of tree removal is higher). The pattern of results is similar to those in the initial eradication scenario with no hurricane, but the numbers are larger. If there was no eradication in 2011, eradication in 2016 yields large gains to U.S. producers that are more than offset by large losses to U.S. consumers and taxpayers, resulting in national losses. Comparing columns (4) and (5), the entries in column (5) are roughly half the size of their counterparts in column (4). This comparison shows that eradication in 2011 reduces by half the U.S. producer gains and U.S. consumer and taxpayer losses, and net national losses from eradication in 2016.

CONCLUSION

This preliminary analysis has significantly extended previous models, to incorporate the dynamics of supply response to prices and policies and to allow for the spatial spread of the citrus canker as a consequence of a hurricane. This analysis has yielded some interesting insights into the role of external shocks that encourage the proliferation of the exotic pest, such as hurricanes, for the economics of alternative control or eradication policies.

First, we explored the benefits and costs of an initial outbreak and an eradication policy in the absence of hurricanes. These simulations indicate that, given the parameters of the industry, its markets and the disease, Florida orange producers would benefit from eradication of the pest in Florida, but U.S. consumers, taxpayers, the nation and the world as a whole would lose. Next we explored the implications of a subsequent hurricane that results in the spread of the disease, and a further policy decision about whether to adopt an eradication program. Given a subsequent hurricane that reintroduces and spreads the disease, the costs and benefits from the initial eradication are lower, a result of the reintroduction of the pest effectively eliminating the stream of benefits and costs from the initial eradication that would otherwise have continued through to the end of the simulation period. The benefits and costs from the subsequent eradication following the hurricane are substantially higher, because the infection following the hurricane extends to several regions whereas the initial infection was confined to the Central region.

An interesting finding is that the interests of Florida orange growers are directly opposite those of the nation as a whole. Suppose the objective is to maximize national surplus. Looking at the scenarios in 2016, it would not pay the nation to eradicate in 2106 regardless of whether it had eradicated in 2011. The optimal strategy in 2011 is also not to eradicate (since eradication in 2011 involves a net social loss if we will not also eradicate in 2016). In contrast, producers would prefer to eradicate in 2016 but not to eradicate in 2011. In a political economy setting, if the government expects that it will have to introduce an eradication program following a hurricane, it would do better also to introduce one in 2011. In contrast, producers would prefer not to eradicate in 2011 if they can be confident of a program being introduced in 2016.

NOTES

[1] Albert K. A. Acquaye is a project economist, Julian M. Alston is a professor, Hyunok Lee is a research economist, and Daniel A. Sumner is the Frank H. Buck Jr. Professor and Director of the University of California Agricultural Issues Center. The authors are all members of the Giannini Foundation of Agricultural Economics, and are employed in the Department of Agricultural and Resource Economics, University of California, Davis.

REFERENCES

Acquaye, A.K.A., Alston, J.M., Lee, H.O., et al., 2005. Economic consequences of invasive species policies in the presence of commodity programs: theory and application to citrus canker. *Review of Agricultural Economics,* 27 (3), 498-504.

Brown, M. and Stover, E., 2002. *Projecting effects of citrus tristeza virus on Florida citrus production.* Department of Food and Resource Economics, Florida Cooperative Extension Service, Institute of Food and Agricultural Sciences, University of Florida, Gainesville. EDIS Document no. FE 330. [http://edis.ifas.ufl.edu/pdffiles/FE/FE33000.pdf]

Gray, R.S., Sumner, D.A., Alston, J.M., et al., 2005. *Economic impacts of mandated grading and quality assurance: ex ante analysis of the federal marketing order for California pistachios.* Giannini Foundation of Agricultural Economics, Oakland. Giannini Foundation Monograph Series no. 46. [http://giannini.ucop.edu/Monographs/46_Pistachios.pdf]

Jetter, K.M., Civerolo, E.L. and Sumner, D.A., 2003. Ex-ante economics of exotic disease policy: citrus canker in California. *In:* Sumner, D.A. ed. *Exotic pests and diseases: biology, economics and public policy for biosecurity.* Iowa State Press, Ames, 121-149.

US International Trade Commission, 2005. *Frozen concentrated orange juice from Brazil. Investigation no. 731-TA-326 (second review).* United States International Trade Commission, Washington. Publication no. 3760. [http://hotdocs.usitc.gov/docs/pubs/701_731/pub3760.pdf]

METHODS
FOR MODELLING
NON-MONETARY
IMPACTS
OF PHYTOSANITARY
POLICIES

CHAPTER 8

ESTIMATING THE ECONOMIC VALUE OF TREES AT RISK FROM A QUARANTINE DISEASE

FRANCISCO J. AREAL AND ALAN MACLEOD

Central Science Laboratory, Sand Hutton, York, YO41 1LZ, UK.
E-mail: f.areal@csl.gov.uk

Abstract. The total economic value of tree species susceptible to *Phytophthora ramorum*, the causative agent of sudden oak death, was investigated in North Yorkshire. The results of a dichotomous-choice contingent-valuation study, using a 'follow-up' dichotomous-choice question, are presented. Two approaches were used in order to obtain the mean willingness to pay (WTP): a bivariate probit model that provides information about the crucial variables that affect the WTP, and the maximization of a log-likelihood function that accounts for a double-bounded bid. Previous studies suggest that the second approach produces more accurate estimates. Using both methods the mean WTP was estimated to be approximately £55 per annum per individual taxpayer over five years. This is similar to values placed by the public on trees susceptible to *P. ramorum* in California, USA.

Keywords: bivariate probit model; contingent valuation; double-bounded model; *Phytophthora ramorum*; sudden oak death; total economic value; willingness to pay

INTRODUCTION

When designing phytosanitary measures to protect plants, it is useful to know the value of the plants that are being protected. Unlike a conventional crop, the valuation of trees in the wider environment requires an understanding that they form part of the rural landscape and therefore represent a 'public good'. The total economic value of an environmental public good includes factors such as how it is used, e.g., for recreation; the public knowing that it exists and will continue to exist; the public's willingness to pay (WTP) for future availability of the resource; and their WTP to avoid an irreversible loss of the resource. There are a variety of techniques available to assess the economic value of environmental public goods (Freeman III 1993). However, few studies have ever quantitavely measured the value of environmental public goods that specific phytosanitary measures are designed to protect. The revised International Standards for Phytosanitary Measures, ISPM 11 (FAO 2003) recognizes that different methodologies can be used to value environmental goods and recognizes the distinction between use and non-use values.

A.G.J.M. Oude Lansink (ed.), New Approaches to the Economics of Plant Health, 119–130.
© 2007 *Springer.*

However, there is little guidance provided on any specific methods available for those using ISPM 11 (Baker and MacLeod 2005).

Contingent valuation (CV) has become one of the most widely used techniques to value environmental public goods. CV refers to approaches based on surveying a sample of a population and applying econometric analysis to the data obtained from the survey to determine what value the population places, or what maximum amount it is WTP, for an environmental public good, or to prevent a specific change in an environmental quality such as loss of trees. Referendum-type questions, with a "yes" or "no" answer can also form part of CV studies with statistical efficiency obtained using a second, follow-up, referendum-type question (Hanemann et al. 1991). With regard to trees throughout the countryside, CV is an appropriate technique because it enables both use and existence values to be measured, which other techniques such as the travel cost method are not able to do. Bräuer (2003) provides an overview of CV for non-economists.

Phytophthora ramorum is a quarantine plant pathogen first identified in the UK in 2002 (Lane et al. 2003). It has been found to be widespread with a low incidence on a range of propagated hosts such as *Camellia*, *Rhododendron* and *Viburnum* (Sansford et al. 2003). A number of common tree species, such as American red oak (*Quercus rubra*), Douglas fir (*Pseudotsuga menziesii*), European beech (*Fagus sylvatica*), Lawson's cypress (*Chamaecyparis lawsoniana*), Sitka spruce (*Picea sitchensis*) and sweet chestnut (*Castanea sativa*) are 'more susceptible' potential hosts. 'Less susceptible' potential hosts include English oak (*Quercus robur*) and *Q. petraea*, European birch (*Betula pubescens*), horse chestnut (*Aesculus hippocastanum*), sycamore (*Acer pseudoplatanus*), European alder (*Alnus glutinosa*) and yew (*Taxus baccata*).

This paper describes the use of contingent valuation to estimate the economic value that a sample of people in North Yorkshire, UK, place on trees that are susceptible to *P. ramorum*, the causal agent of the diseases commonly referred to as Sudden oak death, Ramorum shoot dieback and Ramorum leaf blight (Hansen et al. 2002).

METHODS

All staff from Central Science Laboratory (scientific and non-scientific) were invited to participate in a survey which was conducted in a lecture theatre where they were introduced to the subject of quarantine pests and diseases that could harm trees in the UK. Photographic images of trees, including oak species, in the English countryside were shown, e.g. Figure 1, followed by images from the USA of dead and dying trees infected with *P. ramorum*. Digitally manipulated images were presented showing views of the English countryside with dead and dying trees, supposed to be infected with *P. ramorum*. Although susceptibility to infection may not be a good indicator of potential tree mortality due to *P. ramorum*, we assumed that *P. ramorum* could kill the 'more susceptible' hosts and that infected or dead trees and susceptible trees close by were felled and removed from woodlands to control the spread of the disease. Thus images of the landscapes were presented with the trees digitally removed to simulate the landscape under a scenario where the disease management policy involved cutting and removing infected trees and susceptible trees close by, e.g. Figure 2. There then followed a questionnaire consisting of 22 socioeconomic questions. Respondents were reminded of the social benefits that trees provide. To inhibit respondents from overestimating their WTP (Arrow et al. 1993) they were reminded that other government environmental protection programmes would continue independently.

Figure 1. One of the images of the English countryside used in the study showing tree species susceptible to P. ramorum

Figure 2. Digital manipulation of Plate 1 to illustrate how the landscape may alter if infected and susceptible trees close by were removed to prevent spread of P. ramorum (Trees have been digitally removed from a section of hedgerow in the middle of the left hand side of the view)

The WTP question concerning how much extra tax the individual was willing to pay to protect trees susceptible to P. ramorum was "If there were a public referendum to decide whether a Government Prevention Programme to prevent Sudden oak death from spreading should be implemented, and it cost you £x per year in additional taxes for 5 years, would you vote in favour of it?" In any questionnaire, x was either £30, £50 or £70. These values were uniformly and randomly distributed across respondents. Those who answered that they would be willing to pay £x were then asked if they would pay £20 more, i.e. either £50, £70 or £90, respectively. Those who answered that they would not be willing to pay £x were asked if they would pay £20 less, i.e. either £10, £30 or £50, respectively. The options for the value of £x (the bid amounts) were based on an earlier trial at CSL designed to determine the approximate limits for £x. Double "yes" and double "no" responses were investigated further in order to find protest responses or altruistic motives in the WTP responses. A 'warm glow' effect appears when, in spite of gaining utility from increasing the public-good supply, respondents also gain utility from the act of giving and may be present in the majority of bids giving an overestimation of WTP (Andreoni 1989). Consequently, respondents were asked to explain why they answered "yes" or "no" twice. Reasons that provided evidence for protest or altruistic votes were excluded from determining the mean WTP. For instance, responses such as "I would get pleasure from knowing that I had contributed to a good cause" or "I am opposed to paying for more government programs" were considered not valid.

A probit model is a nonlinear model for estimating values with a binary dependent variable, e.g. the "yes" or "no" responses to the WTP question. We assumed that WTP is distributed in the population according to both a normal cumulative distribution function (cdf) when the bivariate probit model is used, and a logistic distribution function (ldf) when maximizing the double-bounded log-likelihood function. The standard normal cdf is very similar to the ldf and essentially provides identical results. These functions can only be distinguished in very large samples (Aldrich and Nelson 1984).

The probit model is applicable to CV studies with one dichotomous-choice question but by introducing a follow-up dichotomous-choice question, the statistical efficiency improves by the application of a bivariate probit model (Carson et al. 1986).

We adopted the bivariate approach proposed by Cameron and Quiggin (1994), where the two discrete-choice responses are simultaneously modelled as single-bounded, i.e. two correlated WTP equations with jointly distributed normal error terms. This model provides information on what variables are crucial for each of the responses to the WTP question. Moreover, mean WTP for the first and the second question can be calculated from the coefficients obtained from the model. Despite there not being a strong correlation between the two discrete responses in the dataset used, estimation of the mean WTP is feasible using the bivariate probit CV model since bivariate normal probability density functions allow for a zero and non-zero correlation (Cameron and Quiggin 1994). Therefore estimation of the coefficients can be done using a bivariate probit model that would include two related models:

$$Y*_1 = \alpha_1 + \beta_1 B_1 + \sum_{i=2}^{n} \beta_i x_i + \varepsilon_1 \qquad (1a)$$

$$Y*_2 = \alpha_2 + \beta_1 B_2 \sum_{j=2}^{m} \beta_j x_j + \varepsilon_2 \qquad (1b)$$

$$corr[\varepsilon_1, \varepsilon_2] = \rho$$

where Y_1 and Y_2 are the binary responses to the WTP questions; B_1 and B_2 are the bids in the first and second bid question; x_i represent socioeconomic variables and

α's and β's are the coefficients to be estimated. The explanatory variables of model 1 can be different from the explanatory variables of model 2.

Another way of estimating the parameters is by maximizing the following log-likelihood function:

$$\sum_j \sum_i I_{ji}^1 I_{ji}^2 \ln[1 - \psi(f(B^u, x_{ji}))] +$$

$$+ \sum_j \sum_i I_{ji}^1 (1 - I_{ji}^2) \ln[\psi(f(B^u, x_{ji})) - \psi(f(B, x_{ji}))] +$$

$$+ \sum_j \sum_i (1 - I_{ji}^1)(1 - I_{ji}^2) \ln[\psi(f(B^l, x_{ji}))] + \quad (2)$$

$$+ \sum_j \sum_i (1 - I_{ji}^1) I_{ji}^2 \ln[\psi(f(B, x_{ji})) - \psi(f(B^l, x_{ji}))],$$

$$i = 1..N, j = 1..M$$

where I_{ji}^1 indicates a first positive response and I_{ji}^2 indicates a second positive response; B^u and B^l are the upper and lower bid bounds; ψ is the distributed i.i.d. logistically, and f is a function that depends on the bid (B) and a set of socioeconomic variables (x). In respect to the variables included in this model only the first and second bid were included as explanatory variables. The mean WTP, which is shown in Table 5, is what in common usage would be termed the average WTP of the sample.

Whilst the bivariate probit model can be used to determine the mean WTP, previous studies suggest that maximization of a log-likelihood function that accounts for a double-bounded bid produces more accurate estimates of mean WTP. Hence such a technique was used to estimate an alternative mean WTP following Hanemann et al. (1991).

RESULTS

A total of 81 observations were collected. Eighteen (22.2%) were removed from further analysis since they indicated protest or altruistic votes and therefore were not truly valuing the plant protection programme[1]. Additionally, 22 respondents (27.2%) chose the "no answer" option given in the questionnaire[2]. These responses indicate that the 22 took the commitment to pay seriously. From those who responded "no", 77% did not respond because they needed more information; 9% considered other problems are more important than *P. ramorum*; 9% responded by complaining about taxes being the method of payment; and 5% did not believe that trees would be killed by *P. ramorum* in the UK. In this respect, a higher rate of valid responses could be achieved by including more information about the disease and the protection programme. A total of 49.4% of observations was considered non-valid for the aim of valuing the plant protection programme. The probability of answering "yes" twice decreased when this bid amount increased. Conversely the probability of answering "no" twice increased when the bid amount increased (Table 1).

Table 1. *Descriptive statistics (n=41)*

		Responses to the first/second bid [Y = Yes, N = No]			
Bid value thresholds (1st, 2nd)	n	Y/Y(%)	Y/N(%)	N/Y(%)	N/N(%)
£30 (50/10)	13	6 (46%)	4 (38%)	2 (15%)	0 (0%)
£50 (70/30)	12	4 (33%)	5 (42%)	2 (17%)	1 (8%)
£70 (90/50)	14	2 (14%)	6 (43%)	2 (14%)	4 (29%)

Table 2. *Variable names used in bivariate probit models (see Table 3)*

Variable name	Meaning
BID 1	Amount of money asked in the first question
BID 2	Amount of money asked in the second question
MEMENV	Member of an environmental organization (YES=1)
GENDER	Female=1
AGE1829	Respondent's age is in the range of 18-29 year old
BAND	Respondent's staff grade at CSL (proxy variable for income)
DEPEND	Respondent's number of dependents
SODAW	Aware of *Phytophthora ramorum* in the UK
HIGH EDUCATION	First or higher degree

The variable names used in the model and their meanings are shown in Table 2. A total of 6 bivariate probit models were tested and compared in order to choose the best fit using the Log-likelihood test[3]. Table 3 shows the results for each model. Models 2, 4 and 6 are a variation of models 1, 3 and 5, respectively. The only difference is that variables BID1 and BID 2 are transformed to logarithms. Results show that this transformation does not improve the model results except for model 2 where the results are slightly improved and the standard error of the estimated mean WTP for the second question is smaller. Models 3 and 4 added gender and being a member of an environmental organization (GENDER and MEMENV) to the first equation and the proxy variable for salary (BAND) and dependent family members (DEPEND) to the second equation.

Table 3. *Comparison of results of Bivariate probit models*

Parameter	Model 1	Model 2	Model 3	Model 4	Model 5	Model 6
Equation 1						
Constant	1.92**	5.57**	2.59**	6.42**	3.03**	6.74*
	(0.75)	(2.48)	(0.80)	(3.06)	(1.22)	(3.54)
BID 1	-0.03**	-	-0.03*	-	-0.03*	-
	(0.01)	-	(0.02)	-	(0.02)	-
log(BID1)	-	-1.29**	-	-1.42*	-	-1.37
	-	(0.63)	-	(0.84)	-	(0.9)
MEMENV	-	-	0.78	0.75	0.42	0.40
	-	-	(0.87)	(0.88)	(0.97)	(0.96)
GENDER	-	-	-1.08**	-1.06*	-1.17*	-1.16*
	-	-	(0.54)	(0.56)	(0.67)	(0.68)
AGE1829	-	-	-	-	0.86	0.87
	-	-	-	-	(0.75)	(0.76)
WTP	£72***	£74***	£81***	£93	£97***	£137
	(12.54)	(18.93)	(25.58)	(58.84)	(37.36)	(123.47)
C.I. (90%)	£56-88	£49-99	£48-114	-	£48-146	-
Equation 2						
Constant	3.17**	13.51**	1.40	11.02*	1.64	11.72*
	(1.43)	(5.70)	(1.56)	(6.32)	(1.63)	(6.46)
BID 2	-0.06**	-	-0.05**	-	-0.05**	-
	(0.02)	-	(0.03)	-	(0.03)	-
log(BID2)	-	-3.38**	-	-3.13**	-	-3.30**
	-	(1.41)	-	(1.58)	-	(1.59)
BAND	-	-	0.55**	0.57**	0.55**	0.56**
	-	-	(0.24)	(0.24)	(0.24)	(0.25)
DEPEND	-	-	-0.42	-0.41	-0.44	-0.43
	-	-	(0.31)	(0.30)	(0.30)	(0.31)
WTP	£56***	£54***	£57***	£56***	£57***	£56***
	(3.92)	(3.62)	(5.22)	(4.78)	(5.11)	(4.81)
C.I. (90%)	£51-61	£49-59	£50-64	£49-62	£50-64	£50-62
log *L*	-43.49	-43.35	-34.93	-35.12	-33.69	-33.81
Wald Statistic	226.62	231.70	149.77	150.69	139.79	138.19
RHO (p-value)	0.40	0.26	0.52	0.63	0.67	0.74
Pseudo-R^2 statistic	0.14	0.15	0.31	0.31	0.34	0.33

*Significant at 10%   **Significant at 5%   ***Significant at 1%

The addition of these variables improves the model significantly (see Log *L* and Pseudo-R^2)[3]. Therefore, the bivariate probit model 3 shows that including GENDER, MEMENV, BAND, and DEPEND in the model increase the model performance. The likelihood ratio was used to test the join significance[4] with the result of rejecting the null hypothesis of not-join significance of these four added variables (p-value < 0.01). GENDER and MEMENV are crucial in the first response although only GENDER is individually statistically significant. For the second response BAND and DEPEND are crucial although only BAND is statistically significant.

Despite adding AGE1829, models 5 and 6 slightly appear to improve the overall model results (see Log-L), the likelihood ratio statistic could not reject the null hypothesis of not-join significance. This result shows that adding AGE1829 to the model 3 does not improve it enough to include it (p-value>0.10). Moreover, estimates for the mean WTP become less reliable, especially for the first question.

Model 3 provides information about the behaviour of the variables. Thus, a significant negative relationship is found between BID1 and WTP (p-value <0.10) and between BID2 and WTP (p-value <0.05), i.e. the higher the tax the lower probability of answering yes to the WTP question. The fact that the coefficients for BID1 and BID2 are negative and BAND is positive, i.e. the higher the income the higher the probability of answering yes to the WTP question, validates the model in accordance with theoretical expectations. Figure 3 shows the proportion of respondents confirming that they would be willing to pay either of two amounts, the second amount being £20 higher than the first. As expected, the proportion willing to pay decreased as the bid amount increased. For those respondents not willing to pay the first bid, the proportion willing to pay a lower second bid increases as the first bid increases (Figure 4). In addition, gender was also found to be a statistically significant variable. Women are expected to be less WTP than men although no significant correlation between GENDER and BAND, the proxy for salary, was found. Despite neither MEMENV and DEPEND being individually significant, both were found to improve the overall model when included in the first and second equation, respectively. Estimates of WTP obtained from equation 1 are less significant than the estimates for equation 2. This is consistent with the Discovered

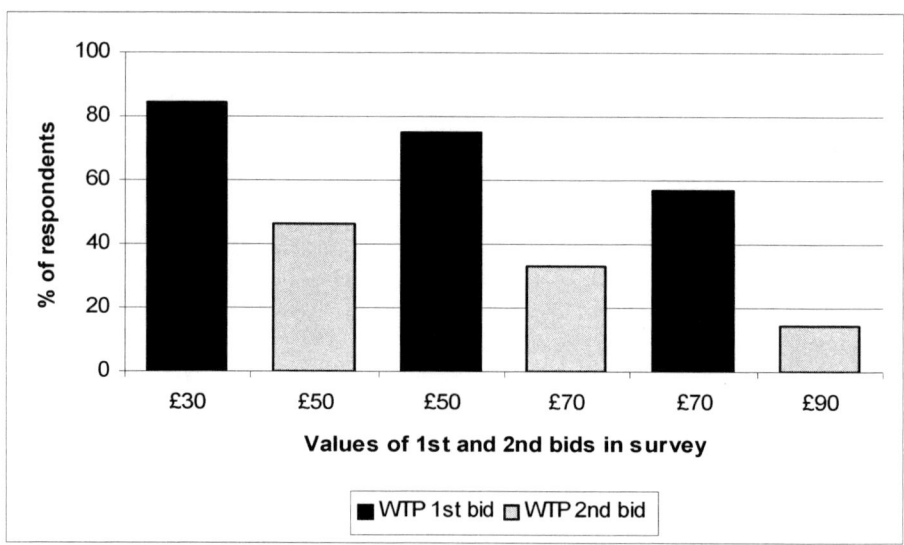

Figure 3. The percentage of respondents willing to pay, both the first and second bid choice values (n= 41)

Preference Hypothesis (DPH) proposed by Plott (1996), which points out that preference consistency is more likely to be observed after repeated valuation trials. Other variables were not included due to correlation problems. Thus, SODAW has been found to be positively correlated with BAND (p-value<0.01), which means that the higher the staff grade the more aware of the disease. Consequently, SODAW was not included in the model because of multicollinearity. BAND was also correlated with HIGH EDUCATION (i.e. first degree or higher degree) (p-value<0.05), which was expected.

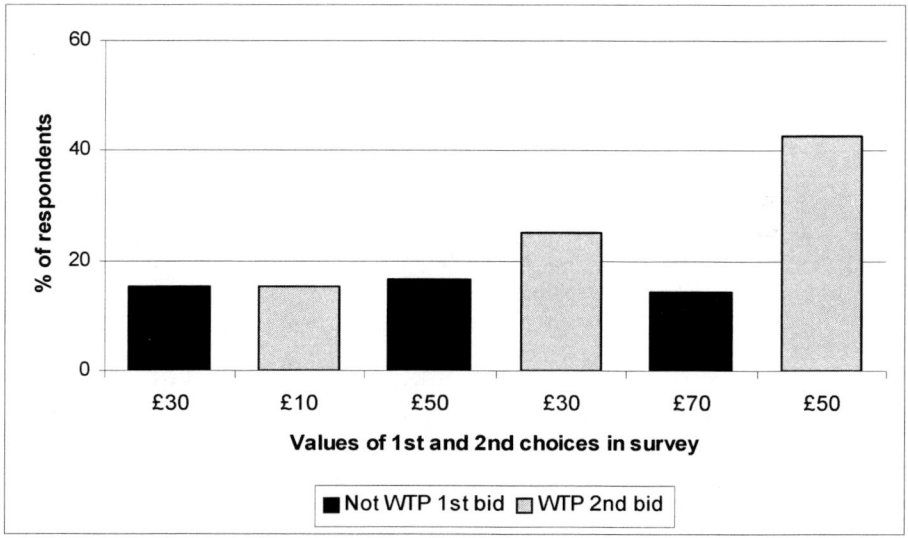

Figure 4. *The percentage of respondents not willing to pay, the first bid but willing to pay the second choice bid (n=41)*

Table 4 illustrates how average WTP is affected by variations in sociological factors such as gender and being a member of an environmental organization.

Table 4. *Comparison of significant groups' first bid Willingness To Pay using model 3*

Sociological factor	mean WTP	Conf. Int. (90%)
Average	£81 (25.58)	£48-114
Female	£47 (10.49)	£33-61
Memenv	£105 (28.39)	£68-142
Female and Memenv	£71 (21.69)	£43-99

Significant at 1%

WTP was estimated using the bivariate probit model and the maximization of the double bounded log-likelihood function. Greater statistical significance was obtained using the second approach (Table 5). The mean WTP, which is shown in Table 5, is what in common usage would be termed the average WTP of the sample.

Table 5. Willingness to pay – result from maximization of the double-bounded log-likelihood function.

Variable	Mean Willingness to Pay	Conf. Int. (90%)
Sample population mean	£55	£51-59

Significant at 1%

DISCUSSION

In the USA, Thompson et al. (2002) also used a double-bounded dichotomous-CV method and estimated that individuals placed a value of between $75 and $83 (approximately £43 and £47) on programmes to preserve oak woodlands in a particular Californian county. Such estimates are similar to our estimate of £55 for an individual's total economic value (TEV) of trees susceptible to *P. ramorum* in the UK. Multiplying up the public's individual mean WTP we have estimated that nationally susceptible trees may have a value of approximately £1.9 billion to the public. Scarpa (2003) collected primary data to augment data from 1992 and estimated the total recreational benefit (use value) of woodlands in the UK to be between £574 million and £962 million. His estimate did not consider non-use values and included all trees in woodlands, not just trees susceptible to *P. ramorum*; yet our estimate is significantly greater. This is probably because, by using CV, the present study includes non-use values, thus increasing the value of trees over their value when only use values are considered. Although the results for TEV may be overestimated due to a small sample population with higher average income than the UK average, and that 90.2% of the sample population were already aware of *P. ramorum*, it appears there is a significant TEV for trees susceptible to *P. ramorum* in the UK.

The relationship between WTP estimated from CV studies and actual observed behaviour has been empirically studied, and investigations have shown that CV performs reasonably well, with a level of accuracy consistent with other techniques used in economics (Cummings et al. 1986; Walsh et al. 1989). Nevertheless, a future wider study across the UK could be required to test the conclusions and WTP estimates obtained by this pilot study.

The model and the estimates of WTP obtained during this study provide crucial information for further research and especially for any future questionnaire design. In this respect, the questionnaire should examine causes of different WTP due to gender. This study showed that females are willing to pay less than males for plant protection programmes although there was no correlation between the salary proxy and gender. Future studies should sample from a larger population across a wide geographic population since it is reasonable to postulate that people from different regions may place different values on trees. The geographical distribution of the sample was not analysed due to the small size of the sample and especially to the constrained geographical location of the survey. The 'warm glow' effect and protest responses were found to be present in 19% and 5% of the entire sample, respectively. Consequently extra information related to the possible causes for altruistic preferences should be obtained in further studies. The protest responses were related to taxes being the method of payment. A higher rate of valid responses could be achieved by including more information about the disease and the protection programme, for example the questionnaire could indicate areas of the country with trees at most risk. Estimates of the mean WTP were found to be different between the first and the following-up question. The correlation coefficient Rho indicates that there was no strong correlation between the two discrete responses in the dataset used. The first question produced higher estimates of the mean WTP than the second question. An explanation for these results may be that respondents become more aware of the amount of money they have to pay when the question is asked a second time. This interpretation is based upon both income and number of repondent's dependents that are critical for the explanation of the response to the second question, i.e. respondents consider these variables when they are asked for the second time. This means that WTP estimates for the second question are preferred to the estimates for the first question. This is consistent with research on DPH, which argues that stable and theoretically consistent preferences typify a product of experience gained through practice and repetition (Plott 1996). Results of this study are also consistent with the literature of double-bounded being more efficient statistically than single-bounded (Cameron and Quiggin 1994).

ACKNOWLEDGEMENTS

We wish to thank CSL staff for taking part in this study; Garry Fry for digitally manipulating the images and Riccardo Scarpa for his advice and suggestions. This study was funded by CSL seedcorn project LB53 9002.

NOTES

[1] Follow up questions regarding possible reasons for unwillingness to pay as well as reasons for willingness to pay were included in the questionnaire to determine valid responses.

[2] Respondents were offered a 'no-answer' option as recommended in the literature (Arrow et al. 1993). This is expected to reduce problematic responses (i.e. answering "yes" or "no" without meaning it).

[3] Pseudo R^2 statistic is calculated as: $Pseudo-R^2 = 1 - \dfrac{LL_U}{LL_R}$ where LL_U represents the log-likelihood functions from the full model and LL_U the log-likelihood function for the restricted model.

[4] The likelihood ratio (LR) test is based on the same concept as the F test in the linear model. The LR test is based on the difference in the log-likelihood functions for the unrestricted and restricted models. The likelihood ratio statistic is twice the difference in the log-likelihood functions: $LR = 2(L_{UR} - L_R)$, where L_{UR} is the log-likelihood value for the unrestricted model and L_R is the log-likelihood value for the restricted model. The multiplication by two is needed so that LR has an approximated chi-squared distribution under the null hypothesis of not-join significance (Wooldridge 2000).

REFERENCES

Aldrich, J.H. and Nelson, F.D., 1984. *Linear probability, logit, and probit models*. Sage, Beverley Hills. Quantitative Applications in the Social Sciences no. 45.

Andreoni, J., 1989. Giving with impure altruism: applications to charity and ricardian equivalence. *The Journal of Political Economy*, 97 (6), 1447-1458.

Arrow, K., Solow, R., Portney, P.R., et al., 1993. Report of the NOAA panel on contingent valuation. *Federal Register*, 58 (10), 4601-4614.

Baker, R. and MacLeod, A., 2005. Pest risk assessments: tools, resources and key challenges. *In:* IPPC Secretariat ed. *Identification of risks and management of invasive alien species using the IPPC framework: proceedings of a workshop in Braunschweig, Germany, 22-26 September 2003*. FAO, Rome, 106-109. [http://www.fao.org/docrep/008/y5968e/y5968e0j.htm]

Bräuer, I., 2003. Money as an indicator: to make use of economic evaluation for biodiversity conservation. *Agriculture, Ecosystems and Environment*, 98 (1/3), 483-491.

Cameron, T.A. and Quiggin, J., 1994. Estimation using contingent valuation data from a "dichotomous choice with follow-up" questionnaire. *Journal of Environmental and Economics Management*, 27 (3), 218-234.

Carson, R.T., Hanemann, W.M. and Mitchell, R.C., 1986. *Determining the demand for public goods by simulating referendums at different tax prices*. Manuscript, University of California, Department of Economics, San Diego.

Cummings, R.G., Brookshire, D.S. and Schulze, W.B., 1986. *Valuing public goods: an assessment of the contingent valuation method*. Rowan & Allenhead, Totowa.

FAO, 2003. *Pest risk analysis for quarantine pests, including analysis of environmental risks and living modified organisms*. Secretariat of the International Plant Protection Convention FAO, Rome. ISPM no. 11. [http://www.fao.org/DOCREP/006/Y4837E/Y4837E00.HTM]

Freeman III, A.M., 1993. *The measurement of environmental and resource values: theory and methods*. Resources for the Future, Washington.

Hanemann, M., Loomis, J.B. and Kanninen, B., 1991. Statistical efficiency of double-bounded dichotomous choice contingent valuation. *American Journal of Agricultural Economics*, 73 (4), 1255-1263.

Hansen, E.M., Sutton, W., Parke, J., et al., 2002. *Phytophthora ramorum* and Oregon forest trees: one pathogen, three diseases: poster abstract. *In: Sudden oak death science symposium: the state of our knowledge, Monterey, California, 15-18 December 2002*. [http://danr.ucop.edu/ihrmp/sodsymp/poster/poster07.html]

Lane, C.R., Beales, P.A., Hughes, K.J.D., et al., 2003. First outbreak of *Phytophthora ramorum* in England, on *Viburnum tinus*. *Plant Pathology*, 52 (3), 414.

Plott, C.R., 1996. Rational individual behaviour in markets and social choice processes: the discovered preference hypothesis. *In:* Arrow, K.J., Colombatto, E., Perlman, M., et al. eds. *The rational foundations of economic behaviour: proceedings of the IEA conference, held in Turin, Italy*. MacMillan, Basingstoke, 225-250. IEA Conference no. 114.

Sansford, C.E., Jones, D. and Brasier, C., 2003. *Pest risk analysis of Phytophthora ramorum*. Available: [http://www.defra.gov.uk/planth/pra/sudd.pdf] (12 Oct 2005).

Scarpa, R., 2003. *Social & environmental benefits of forestry phase 2. The recreational value of woodlands: report to the Forestry Commission*. Centre for Research in Environmental Appraisal & Management, University of Newcastle. [http://www.7stanes.gov.uk/website/pdf.nsf/pdf/nmbrecrep.pdf/$FILE/nmbrecrep.pdf]

Thompson, R.P., Noel, J.E. and Cross, S.P., 2002. Oak woodland economics: a contingent valuation of conversion alternatives. *In: Proceedings of the fifth symposium on oak woodland: oaks in California's changing landscape, October 22-25, 2001*. USDA Forest Service, San Diego, 501-510. USDA Forest Service General Technical Report no. PSW-GTR-184. [http://danr.ucop.edu/ihrmp/proceed/thompson.pdf]

Walsh, R.G., Ward, F.A. and Olienyk, J.P., 1989. Recreational demand for trees in national forests. *Journal of Environmental Management*, 28 (3), 255-268.

Wooldridge, J.M., 2000. *Introductory econometrics: a modern approach*. South Western Thomson Learning.

CHAPTER 9

MULTI-CRITERIA DECISION MAKING TO EVALUATE QUARANTINE DISEASE CONTROL STRATEGIES

MONIQUE C.M. MOURITS
AND ALFONS G.J.M. OUDE LANSINK

Business Economics, Social Sciences Group, Wageningen University, Hollandseweg 1, 6706 KN Wageningen, The Netherlands.
E-mail: Monique.Mourits@wur.nl

Abstract. Decision making in controlling quarantine diseases is a complex, conflicting process, characterized by a mixture of epidemiological, economic and social-ethical value judgments. Policy makers have to integrate these aspects in a consistent and transparent manner in their decision making. Multi-Criteria Decision Making (MCDM) is a tool that is capable of supporting this integration. This paper gives a general overview of available MCDM techniques and provides an application to illustrate the potential support of MCDM in choosing the control strategy that best meets all of these conflicting judgments.

In the application, various strategies to control animal quarantine diseases (such as Foot and Mouth Disease (FMD), Classical Swine Fever (CSF) and Avian Influenza (AI)) were ordered according to the preferences of various stakeholders. Considering the similarity in the complexity of controlling quarantine diseases this 'animal' application provides a good illustration of the potential use of the MCDM evaluation technique within plant disease control.

Keywords: preferences; stakeholders; epidemiology; economics; social ethics

INTRODUCTION

Decision making in controlling harmful plant diseases is a complex process involving a large range of stakeholders with different and often conflicting interests. Their views may represent the interests of the farming community, other sectors of the economy, the consumer or the environment. This may create a situation of conflicting interests, as economic motives may prevail in the views of some, while landscape, environmental or human-welfare motives may be prominent in the view of others.

Multi-Criteria Decision Making (MCDM) could support policy makers in choosing the control strategy that best meets all of these conflicting interests. MCDM techniques deal with complex problems that are characterized by any mixture of quantitative and qualitative objectives. It establishes preferences between alternatives to an explicit set of objectives and measurable criteria.

Although it is one of the most frequently applied tools within operations research and management science (Dodgson et al. 2000; Voogd 1982), MCDM methods are hardly applied in the management of quarantine disease control even though it generally improves the quality and transparency of the decision-making process. A first application in the field of animal disease control was applied in 2004 (Huirne et al. 2005). In this study various strategies to control animal quarantine diseases (such as Foot and Mouth Disease (FMD), Classical Swine Fever (CSF) and Avian Influenza (AI)) were ordered according to the preferences of various stakeholders.

With respect to the complexity of controlling diseases there is a great similarity between the plant production system and the animal production system. This paper therefore provides a description of the performed 'animal disease' MCDM analysis to illustrate the potential use of the MCDM evaluation technique within plant disease control.

The remainder of this chapter is organized as follows. Section 2 gives an overview of existing main categories of MCDM approaches. This is followed by an application of one MCDM technique to the problem of controlling contagious animal diseases such as FMD and CSF. The chapter concludes with a discussion.

OVERVIEW OF MAIN CATEGORIES OF MCDM TECHNIQUES

Multi-Criteria Decision Making is by now a well established paradigm in decision sciences. The key characteristic of this paradigm is that a decision maker does not optimize a single defined objective but aims for the achievement of satisfying levels in the goals or seeks an optimal compromise between several, often conflicting objectives (Romero and Rehman 2003). The general purpose of MCDM is to serve as an aid to thinking and decision making, but not to take the decision. MCDM techniques are capable of dealing with complex problems that are characterized by any mixture of quantitative and qualitative objectives. This is done by breaking the problems into more manageable pieces to allow data and judgments to be brought on the pieces. Next, the techniques reassemble the pieces to present a coherent overall picture to decision makers (Voogd 1982).

Several MCDM techniques have been evolved in the literature since the first seminal paper by Charnes et al. (1955). Goal Programming is the oldest and most widely applied technique. The general set up of a Goal Progamming problem is:

$$Min \sum_{i=1}^{n} (n_i + p_i)$$
$$s.t. \quad f_i(x) + n_i - p_i = b_i \qquad (1)$$
$$x \in F$$

where

n_i = negative deviational variable attached to i-th attribute

p_i = positive deviational variable attached to *i-th* attribute
$f_i(x)$ = mathematical expression for the *i-th* attribute
b_i = target set for the *i-th* attribute
x = vector of decision variables
F = feasible set or region satisfying the constraints.

The central idea behind Goal Programming is that instead of optimizing a set of objectives (or attributes), the decision maker sets targets *(b)* for their achievement. Next, a solution is found by minimizing the deviations (i.e. p_i and n_i) from the set of targets. The vector of decision variables is restricted by the feasible set or region *(F)* that still satisfies the constraints.

A second main category of MCDM approaches is Multi-Objective Programming (MOP). The general set up of an MOP problem is:

$$Eff \ Z(x) = [Z_1(x), Z_2(x).., Z_k(x)]$$
$$s.t. \ x \in F$$

(2)

where *Eff* means the search for efficient solutions in a maximizing or a minimizing sense and where k objectives are involved in the search. Each of the objectives is ruled by a function Z_i. The efficient set is generated using any of three methods, i.e. a weighting method, a constraint method and a multi-criteria simplex method (Rehman and Romero 1993). The essential idea of MOP is the simultaneous optimization of several objectives and that the approach yields Pareto-efficient solutions. A solution is Pareto-efficient if another solution cannot improve it without degrading the performance of at least one other objective in the efficient solution.

Finally, a third category of MCDM approaches is based on Multi-Attribute Utility Theory (MAUT). MAUT approaches try to determine a real-value function, i.e. a utility function for a finite set of alternative systems $x^1, x^2,.., x^m$ such that

$$U[Z_1(x^j), Z_2(x^j).., Z_k(x^j)] \succ U[Z_1(x^i),.Z_2(x^i,..,Z_k(x^i)]$$
$$x^j \succ x^i$$

(3)

where $\succ$ indicates preference of system x^j with respect to system x^i. MAUT is based on the assumptions of perfect rationality underlying the classic von Neumann and Morgenstern utility paradigm. A key assumption in the MAUT approaches is the assumption of preferential independence of objectives, meaning that the trade-off between objectives $Z_i(x)$ *and* $Z_j(x)$ is not affected by the level of $Z_k(x)$ and $k \neq i, j$. In many situations, this preferential independence is too strong; this is particularly obvious in the case where interactions between objectives are apparent. MAUT approaches are generally used in situations where the number of alternatives is small and where the assumption of preferential independence is not problematic.

The simplest operational form of MAUT is based on the assumption that all attribute utility functions are linear, so that the total utility function U is a simple weighted sum of the attribute measures. This assumption implies linear indifferences curves, which is unlikely to be realistic for a wide range of attribute measures, but can be a reasonable approximation over a relatively narrow range of measures.

APPLICATION TO CONTAGIOUS ANIMAL DISEASE CONTROL

Background described MCA research

This section presents an application of a MAUT approach based on the assumption of linear indifferences curves to the problem of controlling contagious animal diseases such as Foot and Mouth Disease and Classical Swine Fever. The MAUT application (hereafter referred to as Multi-Criteria Analysis (MCA)) was part of a large EU research project in which the consequences of outbreaks of contagious animal quarantine diseases were evaluated for various EU member states. Within this EU project, member-state-specific demographic, livestock production, epidemiological and economic data were collected. These data were used as inputs in various modelling modules to obtain insight into the epidemiological and economic impact of outbreaks of contagious animal diseases. The results of these modelling studies along with the results of a detailed questionnaire to elicit the preferences of various stakeholders served as inputs of the MCA framework (Huirne et al. 2005).

Steps within MCA

The applied MCA involves eight steps, as represented by Table 1 and described below.

Table 1. *The 8 steps within the applied Multi-Criteria Analysis*

1.	Establish the decision context
2.	Identify the alternatives to be appraised
3.	Identify objectives and criteria
4.	'Scoring'
5.	'Weighting'
6.	Calculate overall value
7.	Examine the results
8.	Sensitivity analysis

Step 1. Establish the decision context
Within this first step the objective of the MCA should be clearly defined along with an identification of the key players or so-called stakeholders; i.e., decision makers as well as people who may be affected by the decision.

MCA is all about multiple conflicting objectives. There are ultimately trade-offs to be made. Nonetheless, in applying MCA it is important to identify a single high-level objective for which there will be sub-objectives. The aim of this MCA is to make best use of data currently available to support the decision on controlling contagious animal diseases like FMD, CSF and AI.

A key player or stakeholder is anyone who can make a useful and significant contribution to the MCA. Stakeholders are chosen to represent all important perspectives on the subject of the analysis. One important perspective in the field of controlling contagious animal diseases is that of the final decision maker and the animal-health authority to whom the person is accountable. Within this analysis the European Chief Veterinary Officers (CVOs) were asked to express these governmental values through questionnaires. The responses were obtained from a written questionnaire, so there was no interaction or exchange of information/experiences between the various participating CVOs. Beside the group of CVOs, two other groups of stakeholders were asked for their judgments, i.e. an agricultural interest group and a non-agricultural interest group.

Step 2. Identify the alternatives to be appraised
The appraised alternatives per contagious animal disease consisted of the default EU measures (viz. stamping out of detected herds and installation of protection and surveillance zones) and one or more of the following additional control measures:

PRE = pre-emptive slaughter of neighbouring farms within a predefined radius around a detected farm. This measure results in a regaining of the disease-free status (or removal of export bans) 3 months after culling the last detected animal.

VAC_kill = suppressive vaccination within a predefined radius around a detected farm. Vaccination is applied as a suppressive measure; all vaccinated animals will therefore be slaughtered as soon as the epidemic is under control. This measure results in a regaining of the disease-free status 3 months after culling the last detected or vaccinated animal.

VAC_live = protective vaccination within a predefined radius around a detected farm. Vaccination is applied as a protective measure; all vaccinated animals will therefore stay on the farm as soon as the epidemic is under control. This measure results in a regaining of the disease-free status 6 months after culling the last detected animal.

Step 3. Identify objectives and criteria
Assessing alternatives requires thought about the consequences of the alternatives, for strictly speaking it is the consequences that are being assessed, not the alternatives themselves. Criteria and sub-criteria or indicators are the measures of performance by which the alternative control strategies are judged. Criteria are specific, measurable objectives. They are children of higher-level parent objectives, who themselves may be the children of even higher-level parent objectives.

This research is centred on 3 high-level objectives or main criteria, viz. epidemiology, economics and social ethics. Each criterion is broken down into lower-level objectives or indicators to facilitate the scoring process. These clusters of indicators are as presented in Table 2.

Table 2. Overview of main criteria and their indicators, along with the preference weights indicated by the CVOs

Main criteria	CVO weight	Cluster of epidemiological indicators	CVO weight
Epidemiology	53	Duration	28
Economics	30	Number of infected herds	25
Social ethics	17	Size of affected region	19
		Number of destroyed animals	12
Cluster of social-ethical indicators	CVO weight	Number of destroyed herds	
			12
Efficacy	18	Number of destroyed non-farm animals	5
Socio-economic factors	11	Cluster of economic indicators	CVO weight
Macro-economic factors	7		
Commercially interested parties	8	Direct farm losses	
			15
Animal health	8	Cons. farm losses in affected region	14
Animal welfare	7	Cons. farm losses outside affected region	10
Tourism	4	Losses of other participants	11
Non-farm animals	3	Losses of non-agricultural sectors	9
Human health	11	Organisation costs	11
Governmental policy	8	Export restrictions for EU markets	12
Natural life cycle	6	Export restrictions for non-EU markets	9
Food source	9	Tax payer	9

In general, criteria and indicators are defined by help of the stakeholders in an iterative way. However, within the scope of this research, it was not possible to conduct such an extensive, iterative process. The definitions of criteria and indicators are therefore based 1) on the results of a former study in which Dutch stakeholders were interviewed by means of a Group Decision Room session to define the criteria by which animal control strategies should be evaluated (Huirne 2002), and 2) on additional expert consultation.

Step 4. 'Scoring'

When determining criterion scores, specific attention should be paid to the measurement scale. A distinction can be made between a quantitative and a qualitative measurement scale. In case of a quantitative scale, the measurement unit is known, i.e. a quantity has been defined as a standard by which the magnitude of differences can be expressed. Examples of quantitative measurement units are animals, farms, days, and so forth.

The measurement unit of a qualitative measurement scale is unknown. Three qualitative measurement scales can be distinguished with the ordinal scale; having the highest information content, as the numbers of this scale give a rank order. An ordinal scale expresses whether a certain choice possibility is worse or better than any other choice possibility; however, it does not say by how much.

Even if the criterion scores have been determined on a quantitative measurement scale for all criteria, these scores are mutually incomparable since most of the measurement units differ from each other. One criterion might be expressed in number of farms, whereas another criterion is measured in days. To make the various criterion scores comparable, it is necessary to transform them into one common measurement unit by taking care that for each criterion the scores will get a range from 0 to 1. This kind of transformation is called standardization. The method of standardization used for the scores in this study can be written as:

$$\text{Standardized score } i = (\text{score } i / \text{maximum score}) \qquad (4)$$

or each score is divided by the highest score of the criterion concerned. An example is given in Table 3.

Table 3. *A numerical example of the method of standardization*

Criterion 'expected length of epidemic'	Alternative			
	A	B	C	D
Score (days)	76	235	178	156
Standardized score	0.32	1.00	0.76	0.66
Directed standardized score	0.68	0.00	0.24	0.34

An issue related to standardization is the issue of the direction of the criterion scores. For some criteria, a higher score implies an improvement, whereas for other criteria a higher score implies a deterioration. The example criterion 'length of epidemic' from Table 3 is an example of the latter. Each standardization should therefore be accompanied by a consideration of the direction of the scores. In this study, the worst criterion score is given a standardized value of 0, whereas the best criterion score has a standardized value of 1.

Criterion scores can be derived in many different ways. In this study all quantitative scores are based on the results of stochastic simulation modelling studies (Huirne et al. 2005). The presented MCA analyses are directed towards the 95-percentile model results, assuming a risk-averse attitude with respect to the contagious animal disease control. The scores of qualitative indicators are obtained by ranking the alternatives per criterion according to their expected effectiveness. These effectiveness rankings are based on the insights obtained from questionnaires, personal interviews and model studies.

Step 5. 'Weighting'
An indicator's weight (as well as a criterion's weight) should depend on the range of difference in the indicator scores and on how much the stakeholders care about the difference. For instance, most stakeholders consider the length of the epidemic an important decision indicator. However, when alternative strategies would result in an expected duration difference of only a few days, length would not longer be an important decision indicator. In this study, stakeholders were asked to express their

judgments (= weights) on grounds of their subjective knowledge on possible ranges of indicator scores.

The weighting factors applied in this study are based on the results of a written questionnaire. By this questionnaire various groups of stakeholders expressed their judgments using comparative rating scales. Stakeholders had to make judgments of each indicator with direct reference to their judgments of the remaining indicators (Churchill 1995), by dividing 100 points per cluster. In this paper, the main emphasis is on the judgments of the CVOs.

Step 6. Combine the weights and scores for each alternative to derive an overall value

There are several methods by which an alternative's performance across indicators can be aggregated to form an overall assessment. Two of the most applied methods are the simple linear additive evaluation method and the concordance analysis method. The simple linear additive evaluation method combines the alternative's values into one overall value by multiplying the value score on each criterion by the weight of that criterion, followed by a summation of all those weighted scores (Dodgson et al. 2000; Voogd 1982). This method is perhaps the simplest and most intuitive of all aggregation methods. However, the method is only suitable to aggregate scores within a corresponding measurement scale (quantitative *or* qualitative). The concordance analysis is an evaluation method in which the alternatives are ranked by means of their pairwise comparisons in relation to the defined criteria (Nijkamp et al. 1990). Due to the pairwise comparisons, this method is able to aggregate quantitative as well as qualitative scores into one overall evaluation value.

By means of the simple linear additive evaluation method, the overall weighted scores of the three main criteria, epidemiology, economics and social ethics, are obtained. In general, the higher the overall value, the better the alternative control strategy scored within the concerned criterion.

However, the performed multi-criteria evaluation is based on criteria that are partially assessed on a quantitative scale as well as partially on a qualitative scale. To account for the specific characteristics of both measurement scales, a mixed data multi-criteria technique is applied to determine an overall score per alternative. In this mixed data evaluation technique, which is a generalized form of the concordance analysis technique, differences in alternatives are expressed in a condensed way by means of paired comparisons. Standardized scores of each indicator are compared in pairs of the evaluated alternatives, resulting in so-called dominance scores. A positive score implies dominance of one strategy in relation to another while a negative value implies submission. A dominance measure of 0 implies an indifference between the compared strategies. By weighting these dominance scores per criteria, overall dominance scores of the three main criteria are obtained.

To compare the outcomes of the quantitative and qualitative dominance scores, the scores of the individual main criteria are standardized into the same unit. In this way the dominance scores of the quantitative criteria, epidemiology and economics, are comparable to the dominance score of the qualitative criterion, social ethics. By weighting these standardized dominance measures with the aggregated weights of the constituent criteria the overall dominance score per alternative is calculated, which represents the degree in which an alternative was better (or worse) than another alternative.

Step 7. Examine the results
The aggregation of the dominance scores of the three main criteria (viz., epidemiology, economics and social ethics) into one overall dominance score per alternative gives an indication of how much an alternative is appreciated over another. These overall dominance scores also determine the overall ordering of the evaluated control strategies.

Step 8. Sensitivity analysis
Sensitivity analysis provides a means of examining the extent to which the relative importance weight of each criterion/indicator makes any difference in the final results. Interest groups often differ in their views of the relative importance of the criteria (or weights) and of some scores, though weights are often the subject of more disagreement than scores. In this study special attention is given to the comparison between the ranking of alternatives based on the preferences expressed by the CVOs and the ranking based on the preferences expressed by the representatives of the general public.

Using the MCA model to examine how ranking of options might change under different weighting systems can show that, for instance, two options always come out best, though their order may shift. If the differences between these best options under different weighting systems are rather small, accepting a second best option can be shown to be associated with little loss of overall benefit.

Results

Weighting factors reflecting preferences of the CVOs
The response rate of the 25 CVOs on the written questionnaire is about 80 % (i.e. 20 questionnaires). The averaged CVO weights for the three main criteria and their clusters of indicators are represented in Table 2.

With respect to the main criteria, the CVOs prefer the epidemiological criterion with an average relative weight of 53 %. Corresponding average weights for the economic and social-ethical main criteria are 30 % and 17%, respectively. Duration of the epidemic (28 %) and the number of infected herds (25 %) are regarded as the two most important epidemiological indicators. Differences between the relative weights of economic indicators are not as profound as the epidemiological indicators. Direct farm losses (15 %) and consequential farm losses in affected region (14 %) are regarded as the two most important economic indicators. Efficacy (18 %) and social-economic factors (12 %) are considered the most important social-ethical indicators (Table 2).

MCA application to evaluate three FMD control alternatives
This section illustrates the overall MCA results based on the evaluation of FMD control alternatives for one of the studied EU member states, characterized as a net importing, densely populated livestock area.

Overall scores of main criteria
By means of the simple linear additive evaluation method, the overall weighted scores of the three main criteria, epidemiology, economics and social ethics are obtained as demonstrated by Table 4. Based on the overall epidemiological score, the Pre strategy is preferred best, followed by the Vac_live strategy. The overall 0 score on the Vac_kill strategy indicates that – compared to the other 2 alternatives – Vac_kill scores worst on all epidemiological indicators. However, the efficiency with which this strategy controls an FMD epidemic is almost equal to the efficiency of the Vac_live strategy. Due to the fact that the vaccinated animals will be killed afterwards, Vac_kill scores worst on all indicators involving number of destroyed herds or animals. These indicators, therefore, do not strictly reflect epidemiological efficiency; they also reflect a social-ethical element.

Table 4. *Overall weighed scores of three evaluated FMD control alternatives per main criterion. Bold printed values reflect alternatives with highest scores (= highest rank)*

Criterion	Control alternative		
	Pre	Vac_live	Vac_kill
Epidemiology	**36**	27	0
Economics	58	53	**63**
Social ethics	21	**55**	33

The ranking of the alternatives based on the economic criterion demonstrates that the Vac_kill strategy is preferred above the others. However, differences in overall economic values among the alternatives are rather small, as reflected by the small difference in overall value between the first and second ranked alternatives (viz. 5 points).

The economic ranking based on the MCA may differ from the economic ranking based on the result of adding all the losses to one overall value. By utilizing subjective weighting factors, the MCA ranking is not only accounting for the size of the losses but also for, for instance, value judgments on topics as 'who is bearing the losses'.

From a social-ethical point of view, the overall score for Vac_live exceeds the other 2 alternatives. With a difference of at least 22 points, Vac_kill is evaluated as the second best option.

Overall strategy value
Table 5 demonstrates the dominance scores of the three main criteria as a result of paired comparisons of the 3 FMD control alternatives. For instance, the fourth

column of the table describes the results of the comparison between the Vac_live strategy and the Vac_kill strategy. As reflected by the positive scores, the Vac_live strategy dominates the Vacc_kill strategy on 2 of the 3 main criteria (viz. +5.19 on Epidemiology, +0.73 on Social ethics). However, regarding the Economics criterion, the Vac_live strategy is dominated by the Vac_kill strategy (economic dominance score = -0.57).

Table 5. Criteria of dominance scores of the paired comparisons of the evaluated FMD control alternatives (e.g. EU/Pre = EU strategy compared to the Preventive culling strategy)

Criterion	Pre/V_live	Pre/V_kill1	V_live/Pre	V_live/V_kill	V_kill/Pre	V_kill/V_live
Epidemiology	1.75	6.95	-1.75	5.19	-6.95	-5.19
Economics	0.28	-0.29	-0.28	-0.57	0.29	0.57
Social ethics	-1.12	-0.39	1.12	0.73	0.39	-0.73
Total	0.92	6.26	-0.92	5.35	-6.26	-5.35

According to the total dominance scores the Pre strategy is favoured over the other 2 strategies; i.e. all total paired dominance scores are positive. The dominance difference with respect to the Vac_live strategy is, however, small (0.92). Vac_kill is completely dominated by the other strategies as reflected by its negative total dominance scores.

CONCLUSION AND DISCUSSION

The MCA study on animal disease control

Within the EU project various MCAs were conducted to evaluate the ranking of alternative strategies to control contagious animal diseases like FMD, CSF and AI. All analyses were based on the judgment values of the CVOs. Results showed a general tendency towards the ranking of alternatives, which in most of the cases appeared to be independent of the evaluated disease (see for detailed information Huirne et al. 2005). The general tendency can be described as follows:

- In moderately populated livestock areas, the Vac_live and EU strategies are preferred over the other control strategies.
- In densely populated livestock areas, the Pre strategy is preferred over Vac_live strategy and Vac_kill.

Difference in ranking between clusters of countries, comprising regions with comparable density and/or trade characteristics, are possibly underexposed due to the use of 'average' CVO judgements. Disaggregating the panel of CVOs into subgroups according to the density and trade characteristics of the country represented by the CVOs, followed by an analysis per cluster would provide better insight into the possible presence of alternative rankings.

Individual CVOs or – in general – individual interest groups often differ in their views of the relative importance of the various criteria. Using the MCA framework to examine how ranking of alternatives might change under different preferences or weighting systems can show that, for instance, two alternatives always come out best. Their preference order, however, may differ. If the differences between these best alternatives under different weighting systems are rather small, accepting a second best option can be shown to be associated with little loss of overall benefit, as demonstrated by the following example.

The results of the questionnaire demonstrate variation in preferences among three studied interest groups or stakeholders (viz., CVO group, agricultural interest group and non-agricultural interest group). Table 6 summarizes the indicated preference weights for the main criteria per interest group. This overview stresses the contrast in perspectives of the non-agricultural interest group vis-à-vis the other interest groups.

Table 6. Criterion preference weights (%) per interest group

Interest group	Criterion		
	Epidemiology	Economics	Social ethics
CVO	53	30	17
Agricultural	49	33	18
Non-agricultural	51	15	35

An evaluation of the overall dominance scores based on the preference weights of these individual interest groups enables for an examination of the differences in ranking of alternatives. Table 7 shows for each of the interest groups, the overall scores of AI control alternatives for an exporting, densely populated EU member state. Based on the preferences of the CVO and the agricultural interest groups the Pre strategy is ranked first, followed by the Vac-live strategy as second best alternative. From the non-agricultural point of view, the ranking of these two alternatives is just the opposite. However, differences between first and second best alternatives are rather small. The loss of overall benefit associated with the acceptance of the second best alternative is highest for the non-agricultural interest group (difference of 5.8).

Table 7. Overall dominance scores of AI control alternatives based on the criterion weights of the individual interest groups. Bold printed values reflect alternatives with highest scores (= highest rank)

Interest group	Control alternative			Difference with second best alternative
	Pre	Vac_live	Vac_kill	
CVO	**8.3**	7.4	-15.6	0.9
Agricultural	**8.2**	6.8	-15.0	1.4
Non-agricultural	4.2	**10.0**	-14.2	5.8
Veterinarian	7.4	**8.0**	-15.4	0.6

Generally, when opposing stakeholders discuss alternative options, they quickly focus on their differences of opinions, ignoring the effect of many criteria on which there is an agreement. The MCA technique provides a more balanced approach to ensure that all criteria enter the evaluation, with the result that overall differences are not as great as they seem in an unstructured, face-to-face meeting.

Application of MCA in the field of plant disease control

Comparable to the control of animal diseases, decision making in quarantine plant diseases is also complex and conflicting, due to the involvement of various epidemiological, economic and social-ethical value judgments. Control measures as the use of a particular pesticide may limit the spread of the infectious disease (= epidemiological value), but could also affect the subsistence of harmless organisms (= social-ethical value), result in residues in potential food products (= economic and social-ethical value) or even influence the existence of a whole ecosystem (= social-ethical value). Controlling the disease by a measure as complete destruction of plants and plant products may be very efficient from an epidemiological point of view, but could have serious economic consequences for the affected producers and – depending on the magnitude of the outbreak – even affect the world food supply, resulting in a global social-ethical distress.

Based on the findings within the described study it can be concluded that the MCA technique could be a suitable tool to assist plant control decision making by providing structure to debates, ensuring quality conversations, documenting the process of analysing the decision, separating matters of fact from matters of judgment, making value judgments explicit, bringing judgments about trade-offs between conflicting objectives to the attention of decision makers, creating shared understanding about the issues, generating a sense of common purpose, and gaining agreement.

REFERENCES

Charnes, A., Cooper, W.W. and Ferguson, R.O., 1955. Optimal estimation of executive compensation by lLinear programming. *Management Science*, 1 (2), 138-151.

Churchill, G.A., 1995. *Marketing research: methodological foundations*. 6th edn. Dryden Press, Fort Worth.

Dodgson, J., Spackman, M., Pearman, A.D., et al., 2000. *Multi-criteria analysis manual*. Department of the Environment, Transport and Regions, London. [http://www.communities.gov.uk/pub/252/MulticriteriaanalysismanualPDF1380Kb_id1142252.pdf]

Huirne, R., Van Asseldonk, M., De Jong, M., et al., 2005. *Prevention and control of Foot and Mouth Disease, Classical Swine Fever and Avian Influenza in the European Union: an integrated analysis of epidemiological, economic and social-ethical aspects.* EU-Research Report. [http://www.warmwell.com/04dec18brusselsconf.html]

Huirne, R.B.M. (ed.) 2002. *MKZ: verleden, heden en toekomst: over de preventie en bestrijding van MKZ*. LEI, Den Haag. Rapport / LEI. Domein 6, Beleid no. 6.02.14. [http://www2.lei.wur.nl/publicaties/PDF/2002/6_xxx/6_02_14.pdf]

Nijkamp, P., Rietveld, P. and Voogd, H., 1990. *Multicriteria evaluation in physical planning*. North-Holland, Amsterdam. Contributions to Economic Analysis no. 185.

Rehman, T. and Romero, C., 1993. The application of the MCDM paradigm to the management of agricultural systems: some basic considerations. *Agricultural Systems*, 41 (3), 239-255.

Romero, C. and Rehman, T., 2003. *Multiple criteria analysis for agricultural decisions.* 2nd edn. Elsevier, Amsterdam. Developments in Agricultural Economics no. 11.

Voogd, J.H., 1982. *Multicriteria evaluation for urban and regional planning.* PhD Thesis. TU Eindhoven. [http://alexandria.tue.nl/extra1/PRF4A/8203510.pdf]

COSTS AND BENEFITS
OF PHYTOSANITARY
MEASURES

CHAPTER 10

PHYTOSANITARY MEASURES UNDER UNCERTAINTY

A cost-benefit analysis of the Colorado potato beetle in Finland

JAAKKO HEIKKILÄ AND JUKKA PELTOLA

MTT Agrifood Research Finland, Economic Research, Luutnantintie 13, FI-00410 Helsinki, Finland. E-mail: firstname.surname@mtt.fi

Abstract. We have undertaken a temporal cost–benefit simulation of two policies for controlling an invasive pest – the Colorado potato beetle – in the agricultural network of Finland. The policies assessed are the current policy based on a European Union (EU) protected zone (pre-emptive control) and an alternative policy of giving up the protected zone (reactive control). Besides the natural stochasticity related to agricultural production, we assume that the environmental change affects the dynamics of the problem. This change is included by means of three linear trends: i) local climatic change, ii) regional climatic and production change, and iii) biological change in the pest population. Uncertainty is incorporated through stochastic variables and a sensitivity analysis. The main result is that protection is economically viable, provided that there is some future change and a non-insignificant level of winter survival of the pest population.
Keywords: Colorado beetle; protected zone; invasive alien species; simulation analysis

INTRODUCTION

The environment, natural resources and resource-based production are all affected by invasive alien species. Changes in local climatic conditions and abrupt modifications in agricultural policies together with uncertainty related to stochastic environmental fluctuations make invasive-species policies challenging to design and implement. These changes are often exacerbated by changes in the species' characteristics. It is therefore not surprising that invasive species pose an increasing threat to animal and plant health.

Within the European Union's plant health legislation, protected zones are a regional tool available to account for differences in ecological conditions. The aim of the protected-zone policy is to eradicate and prevent the spread of quarantine organisms if encountered in the zone. Member countries can use this voluntary black-list instrument to protect their production environment against specified invasive plant pests. Related national legislation in Finland obliges individual

A.G.J.M. Oude Lansink (ed.), New Approaches to the Economics of Plant Health, 147–161.
© 2007 *Springer.*

farmers to inform the authorities of any quarantine pest observations and to follow orders from the plant protection authorities regarding eradication of those pests. It also specifies penalties for not following orders and obligations and sets out the rights of producers to compensation for eradication costs as well as for the value of the lost crop.

This protection naturally comes at a cost, including the costs of surveillance, labelling, import restrictions, eradication and post-monitoring. The benefits of not having the pest may outweigh these costs, but this is not inevitable as pointed out, for instance, by Mumford (2002) and MacLeod et al. (2005). The aim of this paper is to evaluate the current policy in Finland on the Colorado potato beetle.

COLORADO POTATO BEETLE

The Colorado potato beetle, *Leptinotarsa decemlineata*, (CPB) is the most destructive insect defoliator of the potato. It is oligophagous, feeding exclusively on Solanaceae and primarily on potato. Although the beetle targets other species such as tomato, egg plant, pepper and tobacco, potato is the main host plant in Finland.

The CPB is established in North America, some Central-American countries, many Asian countries and most European countries (except for Britain, Ireland, Norway, Sweden, Finland and some Spanish and Portuguese islands) (EPPO; European Commission 2000). Its presence in Europe dates back some 80 years. It was introduced from the USA to Bordeaux in France in 1922, from where it rapidly spread throughout Europe, reaching Spain and Germany in the 1930s, Portugal and Poland in the 1940s, Bulgaria in the 1950s and Greece in the 1960s (EPPO).

The first invasion in Finland took place in 1983, but was localized and short-lived. The two main invasions were in 1998 and 2002, with the first confirmed case of winter survival observed in 2004. The time-span of the invasion data is not long, but given this dataset, it seems that the invasion pressure is increasing in both the invasion years (2002 vs. 1998) as well as in the interim years (2003-2005 vs. 1999-2001).

Most of the plots affected in both 1998 and 2002 were situated in south-eastern Finland, suggesting that the beetles had spread from either Russia or Estonia, as depicted in Figure 1. The beetle flies only short distances, but can disperse by means of wind-borne long-distance migration, which seems to be its primary mode of transport to Finland. It can also be carried over large distances in sea water, and in addition, transportation of its host plants in, for instance, trucks and trains provides a third method of dispersal (EPPO).

The CPB protected-zone area represents roughly 30 to 40 % of the total potato production in Finland, and includes Satakunta, Turku, Pirkanmaa, Uusimaa, Häme, Kymi and the Åland Islands. The actions within the protected zone and the eradication measures to be undertaken are specified in Council Directive 2000/29/EC and in Regulation 38/04 of the Finnish Ministry of Agriculture and Forestry. Although the protected zone is only for the given areas, national legislation is applied to the entire country and hence the beetle has to be eradicated wherever encountered.

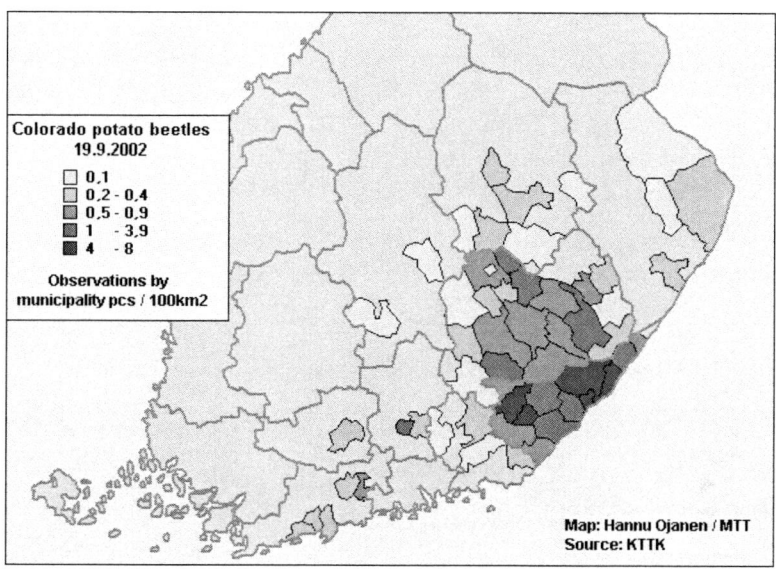

Figure 1. *The density of beetle observations in Finland in 2002*

POLICY COSTS

Cost structure

Costs caused by invasive species may be divided into five categories. In the quantitative analysis carried out in this paper, we include potato production losses, beetle control costs and domestic market effects. In contrast, foreign-trade impacts and environmental, health and cultural costs are excluded from the analysis.

The estimation of costs is affected by natural stochasticities as well as uncertain human behaviour. The physical state of nature itself does not have the main importance in this study. The focus is rather on the economic outcome of that state of nature. Due to the economic focus, also the main uncertainty issues arise from human preferences and decision-making or from the functioning of the society and its institutions. In natural sciences *scientific, stochastic and parametric uncertainties* are important. Related to the CPB, these would translate to uncertainties and natural variation in the invasion process and parameters of the process respectively. These effects are included through stochastic simulation and sensitivity analysis. Given our focus, the main emphasis is on factors affecting human wellbeing such as *impact, policy and value uncertainties*, which here translate to uncertainties in how invasions and policies affect production, and how some unknown economic values affect the process.

The two policies analysed are the current pre-emptive control based on the European Union protected zone and an alternative policy of reactive control by individual producers. In the case of pre-emptive control, the economic cost includes the fixed and variable costs of the protection system. The fixed costs consist of maintaining the appropriate infrastructure and undertaking regular checks to monitor the pest status. The variable costs depend on the invasion magnitude and consist of authority-driven eradication of the pest and financial compensation for the producers.

In the case of a reactive control, two types of costs ensue. First, there are changes in producer surplus due to price changes, pest control costs and the value of lost production, caused by imperfect control or interim damage occurring before control application. Secondly, there may be changes in consumer surplus if the product prices increase due to reduced supply. The costs included in the quantitative analysis of the policies are summarized in Table 1 and discussed in more detail below.

Table 1. Costs of pre-emptive and reactive control

PROTECTED ZONE (PRE-EMPTIVE CONTROL)		NO PROTECTED ZONE (REACTIVE CONTROL)	
Fixed	Variable	Fixed	Variable
Authority fixed costs	Authority variable costs:	No expenses	Changes in surpluses
- fixed inspection points, advertising, telephone, postage, etc.	- inspection visits - area controlled and eradicated - compensation payments		- production losses - control costs - invasion-induced price changes

Costs of pre-emptive control

The actual costs incurred in maintaining the CPB protected zone in Finland, as well as the invasion magnitudes (farms inspected, inspection visits and the number of infestations discovered) in the years 1998–2004 are reported in Table 2.

Table 2. Incurred costs and invasions in Finland in 1998–2004. Note: 'a' denotes a partial estimate

Year	1998	1999	2000	2001	2002	2003	2004
Total cost (€)	N/A	78,712	19,005	45,747	576,371[a]	279,181	29,659
Compensation (cases)	38	11	8	2	85	130	N/A
Compensation (€)	9,340	3,110	3,100	1,850	25,264	31,090	N/A
Farms	400	140	200	200	800	500	238
Visits	500	270	200	240	1,485	773	309
Infestations	149	1	0	2	324	6	29

The fixed costs of the protected zone used in the assessment are estimated from costs incurred in the years 1999–2001. The compensation payments (a variable cost) are subtracted from these costs. The fixed cost thus derived amounts to €37,827/year, which is assumed to include 200 inspection visits per year.

As for the variable costs of protection, a simple model was built to estimate the related variables using the data in Table 2. The estimated costs are as follows: i) inspection cost € 256/visit; ii) control substance cost € 20/ha infested; and iii) eradication cost (including compensation) € 610/ha infested. The model results were then compared with the historical realizations and the model seemed to produce reasonable estimates. As we have no direct data on the hectares invaded, an assumption had to be made that an infested plot is the size of one hectare. Discussions with experts confirm that an average potato plot size of one hectare is not an unreasonable assumption.

In addition, we include the possibility that the protection system may fail in any particular year. In this case, the failed area will be added to the invasion area in the next year. In practice, this is modelled as a product of two variables. The first is the event of protection failure, which is either true or untrue – it either happens or does not happen. If it happens, it will happen in a given percentage of the area invaded in that year. In the present analysis, the failure probability that we use is 0.30, meaning that in every year there is a 30% chance that some beetles will be left unobserved. If there is a failure, then we assume that it will be on 20 % of the invaded area. Thus, protection fails annually on average on 6 % of the invaded area.

In addition, a trend which will increase both of these parameters over time is included in the analysis. This is not a separate trend as such, but is included in all other trends. This is because increasing winter survival, increasing invasion pressure and increasing pesticide resistance (the three trends analysed) all imply that maintaining the protection system will become more difficult, which is then captured in our analysis through increasing failure probability and area.

Costs of reactive control

If the beetle is not eradicated as a part of the protection policy and the producers have to apply control, there will be reactive control costs. These consist of both the cost of the chemical control substances as well as the cost of applying them. The CPB is known not only for its powers of destruction, but also for its ability to rapidly develop resistance to insecticides. For instance, in Russia, Poland and Estonia, the CPB seems to be highly resistant to common pesticides.

The estimates of US chemical-control costs vary widely and have been reported to be US$ 40-$410/ha in Michigan in 1991 (Grafius 1997) and about US$ 300-$700/ha on Long Island due to higher resistance (Raman and Radcliffe 1992). There are no cost estimates available for Europe and, thus, in this analysis, we have applied a non-stochastic figure of € 100/ha for the current analysis. The figure is lower than the costs in the US due to, for instance, a lower level of pesticide resistance in northern Europe. On the other hand, the figure is higher than the cost of € 20/ha used in estimating the costs of the protection system. This is for two reasons. First, the protection system cost does not include work input (which is included in eradication cost category), and secondly, the government agency may have a better knowledge and bargaining power and thus lower-priced control substances than private producers.

Beetle-related variables

Crop damages

The CPB reduces tuber yield of potatoes indirectly by reducing the leaf area, hence decreasing the area available for photosynthesis. The relationship between photosynthetic leaf area reduction and yield loss is not straightforward, but in general reduced leaf area leads to a decreased yield. The relationship is affected, for instance, by how much the leaf area is reduced and at what stage of plant development that is done. The temperature affects the feeding rate of the CPB positively, and also potato types differ in the degree of resistance and damage suffered.

Detailed quantitative descriptions of the beetle's destructiveness in Europe are lacking. EPPO states that in some EPPO countries the yield losses are up to 50 % of the yield. In badly infested areas of Russia, the losses have been reported to be 20 to 70 % of the yield (Parkkonen 2002). The state-wide yield losses in Michigan, USA, are on average 12 % of the yield, although they could be up to 21 % in seriously affected areas (Grafius 1997). These figures may be slightly higher in Europe because most of the beetle's predators, parasites and diseases have remained in America.

Crop damages are modelled as a simple percentage reduction in the yield. Thus, within the area invaded the statistical mean yield is reduced by a given percentage. The estimate should be based on the damages that incur when we have adapted (in the short term) to the presence of the beetle. In a cost–benefit analysis carried out in England (Mumford et al. 2000), it was assumed that when controlled the beetle would impose no damage whatsoever, which we do not find likely. We therefore use a mean of 10 % of the crop for damages by the beetle, and allow this to vary stochastically. The maximum damage is 0.40 and the minimum is zero. In at least 5 % of the iterations the crop damage is zero, and in 5 % of the iterations it is greater than 0.22. The distribution is truncated so that values less than zero are assigned the value zero.

Invasion probability and magnitude

Invasions are modelled as a product of two variables. The first is the invasion event which is either true or untrue with a given probability. We use the figure 0.33, i.e., there will be an invasion on average every three years. If the invasion is true, i.e., if it happens, it will be of a given size. We use a mean of 400 ha, roughly based on the estimated invasion magnitude in the year 2002. The maximum is 935 ha and the minimum is zero. In 5 % of the iterations, the size is below 170 ha and in 5 % it is above 630 ha.

This magnitude is important in two respects. First, in calculating the cost of the protection system, it is the area in which the authorities need to undertake eradication and pay compensation. Secondly, in calculating the costs of reactive control, it is the area on which the beetle produces crop losses, has to be controlled and begins its spread from. Additionally, the invasion magnitude determines the number of inspection visits, so that their number is four times the invasion magnitude, based on past data on the number of infestations and the number of inspection visits. This relationship does not imply causality either way.

Winter survival
The CPB avoids freezing temperatures by digging into the soil to hibernate and by entering a period of diapause, both of which increase its cold tolerance. Its ability for winter survival in Finland is not certain. In the Ukraine, mortality during hibernation has averaged 30 %, but could be up to 83 % (EPPO). In addition to winter temperature, if the summer is too cold, there is no opportunity for proper development. Even a mild winter can then exterminate the population. In Russia, it has been estimated that the requirement for a full generation developing (required for establishment) is at least 60 days of temperature being over +15°C and winter temperature not falling below –8°C (Vlasova, cited in EPPO). Given the recent experiences in Estonia, Russia and Finland, these conclusions may need to be reviewed.

Winter survival in the model affects the spread of the beetle in reactive control, where the protection system is abandoned and coexistence with the beetle becomes reality. It also affects the survival of the population under the protection system when protection has failed in some area. The analysis assumes that in these instances some proportion of the beetle population (or rather, of the area invaded) survives the winter and adds to the invasion area in the following year. The analysis uses a value of 30 % for winter survival. The maximum value is 0.87 and the minimum is zero. In 5 % of the iterations the value is below 0.07 and in 5 % of the iterations it is above 0.53. To anticipate the results, it turns out that this variable is extremely important, and perhaps one for which reducing the uncertainty regarding its true value would be valuable.

Spread variables
In addition to new invasions and the winter survival of the existing populations, the spread of the beetle determines the extent to which the beetle will be present in the country in the event of giving up the protection system. If not controlled, the offspring population of a single CPB pair may become very large. If authority-driven protection is not undertaken (i.e. in reactive control), we assume that there will be some spread already in the first summer. In the case of pre-emptive control, it is assumed that coordinated actions can curb any further spread. In other words, the controlled area is always somewhat smaller under a coordinated authority-driven protected zone than under a control situation which is based on the actions of individual producers. In the latter case, the area controlled is the initial year spread times the initial invasion magnitude. In this analysis, the mean of initial year spread is 1.5. What this means is that if the initial invasion size is 400 ha, then under reactive control the area invaded during the first summer will be 600 ha, while under pre-emptive control it will be 400 ha. The distribution of the variable is restricted to be greater than or equal to 1 and the maximum value is 2.5. In 5 % of the iterations, the value is below 1.13, and in 5 % of the iterations it is above 1.87.

We compare the costs of pre-emptive control with two alternative spread scenarios of reactive control – the first with a logistic spread and the other with a linear spread.

Scenario 1 of reactive control assumes logistic spread. Put simply, the area invaded in year *t+1* is area invaded in year *t* times the spread variable. In reality, also new invasions, winter survival and the extent of the invasion in year *t* affect the spread. The mean of the spread parameter is 1.8 in the analysis. The distribution of the variable is restricted to be greater than or equal to 1 and the maximum value is 4.41. The variance of the variable is assumed to be fairly large. In at least 5 % of the iterations the value is 1, and in 5 % of the iterations it is above 2.84.

Scenario 2 of reactive control assumes linear spread. This means that the beetle will spread to a given area every year, regardless of the area it currently occupies. This area is always non-negative and assumed to be on average the same size as the original invasion, i.e., 400 ha with the maximum size being 860 ha. Ninety percent of the iteration values are located between 235 ha and 564 ha. In addition to stochasticity, the linear-spread area is affected by stochastic winter survival.

LOCAL CHANGE

A further component in the analysis is local change. This materializes through changes in the mean variable values governing the dynamics of the system over time. Three trends are studied, all at three different levels: i) no change; ii) slow change; and iii) rapid change. There are no data describing the dynamics, rather we simulate alternative future scenarios and evaluate subsequent realisations.

Trend 1: Local climatic change (population winter survival)
Through climatic change and changes in the beetle's winter tolerance, it is possible that the winter survival of the beetle population improves (Knight and Wimshurst 2005; Walker and Steffen 1997). In the simulation, the change materializes through increases in the percentage share of those who survive the winter. The winter survival variable is created with a linear trend in the deterministic mean of the variable, but in the analysis, stochastic variation is allowed around this mean. We assume that, in slow change, winter survival increases in 50 years from 30 % to about 45 %. In rapid change, the change is from 30 % to about 60 %.

Trend 2: Regional climatic change (invasion pressure)
Due to regional climatic change, increased trade, modified production practices and northward advancement of the permanent beetle population, it is to be expected that invasions will become more frequent. In the simulations, the invasion probability as well as the average size of an invasion increases over time. We assume that in 50 years the average size of an invasion increases from about 400 ha to about 600 ha in slow change and to about 800 ha in rapid change. The annual invasion probability increases from about 33 % to about 50 % in slow change and to about 65 % in rapid change.

Trend 3: Increasing pesticide resistance
The beetle is capable of quickly developing resistance to different pesticides. Thus, the effectiveness of pesticides decreases and the costs increase over time. In the analysis, the impact of increasing pesticide resistance functions through increasing costs of reactive control as well as the control substance component of the variable costs of protection. We assume that the variable costs of protection increase from € 20/ha to about € 40/ha in slow change and to about € 50/ha in rapid change. In reactive control, the costs increase from € 100/ha to about € 200/ha in slow change and to about € 250/ha in rapid change.

EX-ANTE SIMULATION ANALYSIS

The planning horizon in the ex-ante simulation is 50 years. In this period, invasion events take place randomly. The length of the analysed period is chosen to demonstrate the impact of changes, giving them sufficient time to materialize. The analysis is conducted for 300,000 iterations in order to have a sufficient representation of various variable combinations. We have computed the present values of the policies using a discount rate of 2 %.

Table 3 depicts the number of iterations (cases) in which one of the policies imposes lower costs than the other. For instance, in the case where all trends are slow, in 93.6 % of the iterations pre-emptive control imposes lower costs than reactive control. In other words, in 93.6 % of different realizations of future, pre-emptive control produces positive net benefits.

Table 3. Cases (%) where the one policy imposes lower costs than the other policy

Cases %	Scenario	Pre-emptive control	Reactive control
No trend	Scenario 1	37.5 %	62.5 %
	Scenario 2	47.3 %	52.7 %
Slow trend	Scenario 1	93.6 %	6.4 %
	Scenario 2	67.6 %	32.4 %
Rapid trend	Scenario 1	100.0 %	0.0 %
	Scenario 2	93.1 %	6.9 %

When all trends are off, reactive control is the least-cost policy choice in the majority of cases (62.5 % under Scenario 1 and 52.7 % under Scenario 2). Similarly, when all trends are either slow or rapid, pre-emptive control is the least-cost policy choice in the majority of cases (93.6 % and 100.0 % under Scenario 1 and 67.6 % and 93.1 % under Scenario 2).

The trends thus enhance the profitability of protection. Whenever there is some anticipated change, pre-emptive control is the cost-minimizing strategy in 68-100 % of the cases. This result can also be looked at from the other perspective. Assuming a risk-neutral society and either no change in the future or certain 100 % winter mortality, it would be economically sensible to abandon the protection system. Under such assumptions reactive control would be the least-cost policy choice in 53-63 % of the possible realizations of future.

In all cases, the mean and median costs are very close to each other, indicating that the distribution of costs is fairly balanced. The differences in present value mean cost estimates under pre-emptive control and the two scenarios of reactive control are not very large in the context of no change (€ 8.3, € 8.0 and € 8.3 million, respectively) and to some extent under slow change (€ 13.1, € 17.3 and € 13.7 million, respectively). In the case of rapid change, the differences become larger (€18.9, € 40.0 and € 22.0 million, respectively).

The trends unambiguously increase the mean, minimum and maximum costs of both strategies, but increase the costs of reactive control relatively more. This is also evident from looking at the number of cases where pre-emptive control is cheaper in Table 3. There we already noticed that pre-emptive control becomes more preferred the more change there is. This is because with the increasing trends the pest is able to spread to larger areas and survive the winters better, and becomes more expensive to control. Finally, if there is no winter survival, costs are unambiguously lower with reactive control than with pre-emptive control.

As for the variability of the cost estimates, it is remarkable how the present value of costs varies from the minimum cost of Scenario 1 under no change of less than € 0.9 million (or less than € 0.4 million with no winter survival) to the maximum cost of Scenario 1 under rapid change of nearly € 121 million. The highest possible estimate is thus over 140 times greater than the lowest estimate.

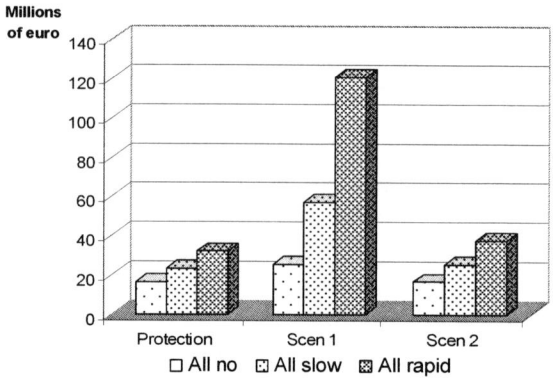

Figure 2. Maximum present-value costs of protection and reactive control (Scenarios 1 and 2)

This result can also be seen by looking at the maximum costs of the policies, as depicted in Figure 2. The maximum costs under rapid change in Scenario 1 can be very much higher than the maximum costs associated with pre-emptive control. Hence, if we are fairly certain that Scenario 1 is the more adequate description of the likely spread of the CPB, then should we choose to abandon protection, the risk from doing so would be very high indeed. However, if we consider Scenario 2 to be a more truthful description (or if we think that there will be no change in the future), there is not so much difference in the risk associated with the two policy options.

Another way to look at the results is to compute the benefit:cost ratios (BCRs) by dividing the benefits of the protection system (i.e. avoided reactive-control costs) by the costs of the protection system (Table 4). BCR denotes *by how much* one of the policies is more economical than the other. Any ratio below 1 implies that protection is more expensive than reactive control, and for instance the mean ratio of 1.32 for slow change under Scenario 1 means that giving up pre-emptive control would on average be 1.32 times more expensive than continuing with it.

Table 4. The BCRs of each strategy and scenario

B:C RATIOS		Scenario 1	Scenario 2
NO CHANGE	Min. | Mean | Max.	0.30 | 0.96 | 2.40	0.39 | 1.00 | 1.90
SLOW CHANGE	Min. | Mean | Max.	0.54 | 1.32 | 3.77	0.57 | 1.06 | 1.75
RAPID CHANGE	Min. | Mean | Max.	0.86 | 2.12 | 7.04	0.67 | 1.17 | 1.90

The minimum BCRs are systematically below 1. Hence, protection cannot be automatically regarded as a dominant least-cost strategy. On the other hand, the maximum BCRs are systematically greater than 1, and therefore by a similar argument reactive control cannot be regarded as a dominant least-cost strategy. Interpretation of results is further complicated by the fact the mean BCRs are at a range of 0.96-2.12, depending on the scenario and the level of change. Hence, the mean BCRs are fairly close to 1 and on either side of it, indicating that the variable values that have been used are such that it cannot be established for certain which policy is the more economical choice.

However, again the trends strengthen the viability of the protection system. The more we expect the climate and the pest to change, the more economical the investment in the protection system becomes. The mean BCRs can be compared to the BCR of 7.5 estimated by Mumford et al. (2000) for the British CPB protected zone.

At an extreme, the protection system is about three times more expensive than reactive control (BCR of 0.30 under Scenario 1 with no change). At the other extreme, reactive control is about seven times more expensive than protection (BCR of 7.04 under Scenario 1 with rapid change). These results again raise the same arguments as those mentioned when the maximum costs of the policies were discussed. Somewhat more interesting is the fact the BCR under Scenario 2 is fairly robust and hardly affected by the level of change, implying that the spread of the beetle is not promoted by change as much under Scenario 2 as is the case under Scenario 1. This is largely due to the fact that spread of the beetle is much more modest under Scenario 2 than under Scenario 1 and, hence, the potential damages are also lower.

Figure 3 plots the cumulative density functions of net benefits under different levels of change in Scenario 1 (panel on the left) and Scenario 2 (panel on the right). The points marked with dashed lines represent the probabilities at which the net benefits of the protection system (cost of reactive control less the cost of

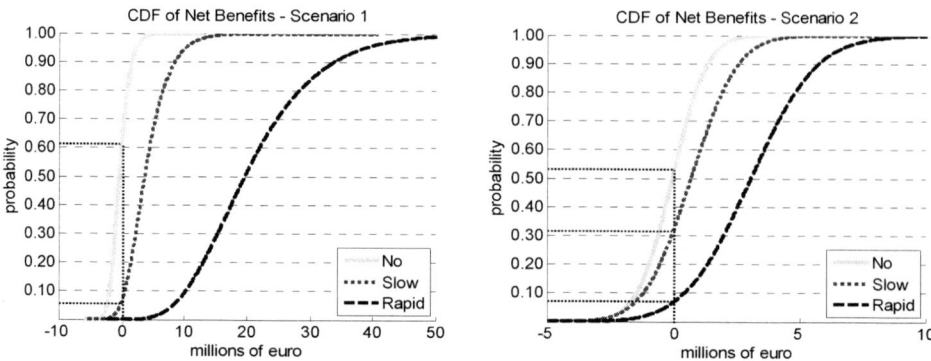

Figure 3. *Cumulative distribution of net benefits of protection in Scenario 1 and Scenario 2*

pre-emptive control) are positive under the two scenarios when subjected to different levels of change. These levels are the same as the percentages reported in Table 3. Figure 3 clarifies how to take into account the level of risk we are willing to accept. For instance, under Scenario 1 there is a 50% probability that the net benefits of protection are negative (no change), less than ca. € 4 million (slow change) or less than about € 20 million (rapid change). Similar assessment can be done for all probabilities and the associated net benefits.

Winter survival

To account for uncertainty, a standard sensitivity analysis with low/high values was carried out. The variable that was found to be most influential was winter survival. Figure 4 represents the impact of different levels of winter survival on the mean BCRs. Winter survival is an important variable especially under Scenario 1, in terms of both mean costs and the mean BCR. It should be noted that, for instance, 100 % winter survival would imply that the BCR is about 30 under Scenario 1 and about 14 under Scenario 2, suggesting very high costs for giving up protection.

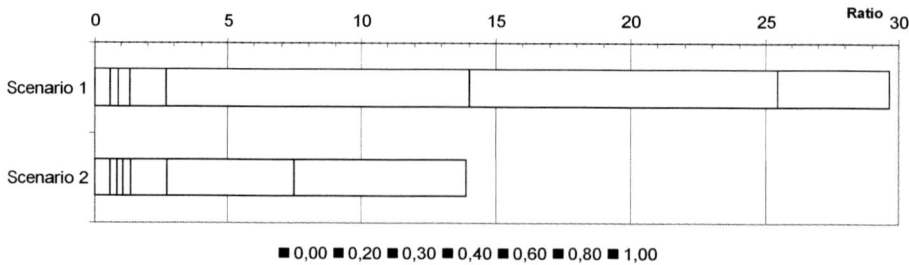

Figure 4. *The mean BCRs with different levels of winter survival*

It is evident that if the level of winter survival is even moderately greater than assumed (say, 40 % instead of 30 %), the mean results are not as ambiguous as before. With 40 % winter survival the mean BCR is greater than 1 under both scenarios, implying that protection is economical. If the level is moderately lower (say, 20 %), the mean BCRs under both scenarios are less than 1, implying that protection is more expensive than reactive control. Furthermore, slightly greater changes in the survival rate (assume, say, 60 % survival) take the mean BCR to 14 under Scenario 1 and to about 3 under Scenario 2. Hence, the importance of this variable is immense, and the implications of the analysis are very much dependent on the value of winter survival that is chosen.

The level of winter survival then naturally affects not only the BCRs but also the mean and maximum costs of the policies. For instance, with 100 % winter survival the present-value mean costs of reactive control under slow change would increase from about € 13-17 million to € 230-490 million. Similarly, the maximum costs would increase from about € 25-57 million to € 410-755 million.

Change through trends

In the basic results, all trends are simultaneously either off, slow or rapid. In the sensitivity section, all trends are set at slow. Let us now have a look at the trends separately. The four different categories of change are: i) local change; ii) regional change; iii) local and regional change; and iv) development of pesticide resistance.

On basis of the analysis, local change is the most important trend. This is consistent with the results of the sensitivity analysis, where it was found out that winter survival is the single most important variable, and it is that same variable that is increasing in local change but not in any other separate trend.

Regional change (increasing invasion pressure) plays a role in increasing the mean present value of all future costs of the policies, but not so much in the relative profitability of different policies (BCRs). The impact of local and regional change combined is similar to the impacts of local change, only with higher magnitude.

Increasing pesticide resistance plays only a minor role, both in terms of impacts on BCRs as well as in the mean present-value costs. It is plausible that the increase in control costs is relatively insignificant when compared to the other policy costs incurred. This result is consistent with the finding that the reactive-control cost is fairly insignificant and can be increased by 50 % without any real influence on the results. Whether its value has been set too low in the analysis is a point of discussion.

Although the trends themselves are plausible and likely in the future, the functional form and the magnitude of the trends are uncertain – they are subject to much scientific uncertainty and call for further research.

DISCUSSION

The Colorado potato beetle is a typical wide-spread plant pest and a nuisance in North America, Europe and to an increasing extent in Asia, affecting productivity of an important food crop. Hence, it is of general interest worldwide. In terms of the European Union, the case is of specific interest as protected zones are an EU-wide instrument that has been designed for protecting plant production. In Finland, the case is of interest because potato is a relatively important national food crop. Furthermore, the CPB provides a convenient case for studying the effects of invasions, uncertainty and local change in fairly manageable circumstances with some data on invasions available and relatively few externalities present.

Given the life-history characteristics of the CPB, there are five important factors to take into account from an economic point of view. First, the beetle has spread very rapidly across the continent, although its spread has slowed down as it has approached its ecological limits. Second, in propitious environmental conditions, its population size can increase extremely rapidly. Third, it is capable of causing significant damage to potato plants. Fourth, cold summers and winters present an obstacle to its establishment, but so far its ability to establish itself permanently in Finland has been difficult to predict. Finally, lack of natural predators and ability to develop resistance to chemical control substances make the beetle difficult and expensive to control

In this analysis, we have concentrated mainly on direct costs and benefits of protection. The general results indicate that protection is economically viable, provided that there will be some future change and non-insignificant level of winter survival of the pest population. Under the conditions and assumptions of this study, we can give up protection if we are certain that there is no future change or that winter survival stays permanently below about 20 %.

The risk associated with giving up protection is, however, much larger than that associated with protection. At the extreme, the cost of giving up protection may be over twenty times greater than continuing with it. Sensitivity analysis conducted for a range of variables reveals that winter survival is the most important variable. Other significant variables include logistic spread rate and the variable cost of protection.

The analysis above is mainly concerned with economic efficiency of the policy concentrating on direct benefits. In a complete analysis, indirect benefits and effectiveness of institutions have to be accounted for. For instance, coordinated protection system versus decentralized decision-making by numerous independent farmers may indirectly affect the outcome through development of resistance or loss of export possibilities. Similarly, social-justice issues need more attention. Imperfect markets mean that changes in domestic supply can have price effects. The economic implications of this come through changes in consumer and producer surplus, and various types of transfer mechanisms can be designed to make sure that certain agents pay for the costs. It is a task of policy makers to decide who pays the costs of the policy and who gets to take part in making the policy. Also, when in time those costs occur and decisions are made is a matter of social justice.

In the course of the analysis, a need for more information has surfaced. Besides the need for natural-science data, the following issues could be of interest when making policy decisions: i) who pays for the policies and when in time do the costs of different policies occur? ii) what are the impacts of possible nonlinearities in costs of prevention and reactive control? iii) what are the impacts of different policies on foreign trade in the form of sanctions, reputation and pesticide use?

iv) how do different reactive-control alternatives rank in terms of economic efficiency? v) what are the implications from the fact that one of the policies (giving up protection) is irreversible, whereas the other one is not? vi) what are the implications from the fact that there are both professional and habitual potato producers, whose behaviours may differ from each other? vii) what are the implications from the fact that the protected zone acts as a buffer zone protecting potentially also Sweden and Norway? viii) if protection is given up at some point in the future, what is the optimal timing for such a switch? ix) what are the lessons learned from the case of the CPB for a more general assessment of invasive plant pests in Finland? and finally x) what is the role of the CPB protection policy in the wider framework of biosecurity measures given limited resources by the state? These issues should be examined in later work.

REFERENCES

EPPO, *Data sheets on quarantine pests: Leptinotarsa decemlineata.* [http://www.eppo.org/ QUARANTINE/insects/Leptinotarsa_decemlineata/LPTNDE_ds.pdf]

European Commission, 2000. EC Council Directive 2000/29/EC of 8 May 2000 on Protective measures against the introduction into the Community of organisms harmful to plants or plant products and against their spread within the Community. *Official Journal of the European Communities,* L 169/1 (10.7.2000), 1-112 and subsequent amendments. [http://europa.eu.int/eur-lex/pri/en/oj/dat/2000/ l_169/l_16920000710en00010112.pdf]

Grafius, E., 1997. Economic impact of insecticide resistance in the Colorado potato beetle(Coleoptera: Chrysomelidae) on the Michigan potato industry. *Journal of Economic Entomology,* 90 (5), 1144-1151.

Knight, B.E.A. and Wimshurst, A.A., 2005. Impact of climate change on the geographical spread of agricultural pests, diseases and weeds. *In:* Alford, D. and Backhaus, G. eds. *Introduction and spread of invasive species:symposium, 9-11 June 2005, Berlin, Germany.* British Crop Protection Enterprises. BCPC Symposium Proceedings no. 81.

MacLeod, A., Baker, R.H.A. and Cannon, R.J.C., 2005. Costs and benefits of European Community (EC) measures against invasive alien species: current and future impacts of *Diabrotica virgifera virgifera* in England & Wales. *In:* Alford, D. and Backhaus, G. eds. *Introduction and spread of invasive species:symposium, 9-11 June 2005, Berlin, Germany.* British Crop Protection Enterprises. BCPC Symposium Proceedings no. 81.

Mumford, J., Temple, M., Quinlan, M., et al., 2000. *Economic policy evaluation of MAFF's plant health programme: report to Ministry of Agriculture Fisheries and Food.* ADAS Consulting, London.

Mumford, J.D., 2002. Economic issues related to quarantine in international trade. *European Review of Agricultural Economics,* 29 (3), 329-348.

Parkkonen, M., 2002. Venäjän pelloilla aletaan viljellä tuholaisia tappavaa perunaa. *Helsingin Sanomat* (26.09.2002), C1.

Raman, K.V. and Radcliffe, E.B., 1992. Pest aspects of potato production. Part 2. Insect pests. *In:* Harris, P.M. ed. *The potato crop: the scientific basis for improvement.* 2nd edn. Chapman & Hall, London, 477-499.

Walker, B. and Steffen, W. (eds.), 1997. *A synthesis of GCTE and related research: the terrestrial biosphere and global change: implications for natural and managed ecosystems.* The International Geosphere-Biosphere Programme, Stockholm. IGBP Science no. 1.

CHAPTER 11

THE BENEFITS AND COSTS OF SPECIFIC PHYTOSANITARY CAMPAIGNS IN THE UK

Examples that illustrate how science and economics support policy decision making

ALAN MACLEOD

Central Science Laboratory, Sand Hutton, York YO41 1LZ, UK.
E-mail: a.macleod@csl.gov.uk

Abstract. Three examples of benefit/cost analyses (BCA) conducted in recent years in the UK to support phytosanitary policy are summarized. Following the first UK outbreak of *Thrips palmi*, the costs incurred during the eradication campaign were compared with potential-losses forecast by modelling the spread and impact of *T. Palmi* in glasshouse crops over ten years. The resultant BCA justified the strict statutory action taken to achieve eradication. The second example, the eradication of a plant pathogen, *Ralstonia solanacearum*, from a river system showed that the expense of a statutory campaign is justified only if eradication can be achieved within a few years. A more protracted campaign would lead to costs outweighing benefits. A third analysis examining the economic impact of implementing EU control measures on *Diabrotica virgifera virgifera*, an insect pest of maize that is currently spreading across Europe, highlights the importance of assessing the cost of implementing measures as well as the benefits of avoiding losses caused by the target pest. This last example shows that strict implementation of control measures can be more costly than the damage likely to be caused by the pest. The strengths and weaknesses of benefit/cost studies, and their future use in relation to plant health issues are discussed.

I am grateful to Prof. Lansink and the Frontis organization for inviting me to participate in the workshop. Benefit/cost analyses were funded by Defra Plant Health Division.
Keywords: benefit/cost analysis; *Diabrotica virgifera virgifera*; economic evaluation; *Thrips palmi*; plant health; potato brown rot; quarantine; *Ralstonia solanacearum*; western corn rootworm

INTRODUCTION

The international trade in plants and plant products acts as the primary mechanism for the unintentional introduction of non-indigenous pests (Levine and D'Antonio 2003). For example, over 80 % of non-indigenous pests that established in the USA between 1980 and 1993 have been assessed as having entered the USA unintentionally through international trade (Jenkins 1999). The World Trade Organization (WTO) Agreement on Sanitary and Phytosanitary Measures (SPS Agreement) allows member countries to protect crops and other plants from the risks

A.G.J.M. Oude Lansink (ed.), New Approaches to the Economics of Plant Health, 163–177.
© *2007 Springer.*

of pest introduction that can arise from international trade by applying protective measures to trade pathways. The SPS Agreement requires that such protective measures be based on risk assessment techniques developed by a relevant international organization. For phytosanitary issues, the International Plant Protection Convention (IPPC) is the relevant organization. In developing a global standard for plant pest risk analysis (FAO 2003), the IPPC recommends that, when selecting appropriate risk management options, those measures with an acceptable benefit to cost ratio should be considered (FAO 2003, part 3.4). Using benefit/cost analysis (BCA) to inform phytosanitary policy is a relatively recent development. Early examples include Rautapaa (1984), who examined the benefits and costs of maintaining Finland free from *Liriomyza trifolii*, the chrysanthemum leaf miner, and Pemberton (1988), who examined the benefits and costs associated with excluding the highly contagious bacterial disease potato ring-rot, caused by *Clavibacter michiganensis* ssp. *sependonicus*, from the UK. Despite these examples, the application of economic analysis to protecting plants from exotic pests was still in the stages of early development towards the end of the 20th century (Krystynak 1991). More recently, Sumner (2003) considered that there was still relatively little economic analysis of government policies related to exotic agricultural pests although he did recognize that where such activity took place the policy evaluation was being conducted at a more rapid pace. Now, when developing phytosanitary policy, governments increasingly require detailed economic impact analyses to inform and help shape quarantine decisions.

This paper provides three examples of summaries of BCA studies that have examined the actual or potential economic impacts of implementing phytosanitary campaigns against quarantine pests in the UK.

EXAMPLE 1: *THRIPS PALMI* – AN INSECT PEST IN GLASSHOUSES

Background and the UK outbreak

Thrips palmi is a polyphagous pest that feeds on the midribs and veins of leaves and stems of more than 50 plant species from over 20 families. Hosts include a wide range of economically important vegetable and ornamental plants. Originating in South East Asia, *T. Palmi* was found in India in the 1960s and has become an increasingly important pest around the world as it has spread within tropical regions of Africa, Australia, South America, Hawaii and the Caribbean, and in sub-tropical regions of Florida and Japan. Since 1978, *T. Palmi* has become the most serious pest of a number of glasshouse and field crops in southern and western Japan, regularly causing crop losses (Kawai 1990). The first European outbreak of *T. Palmi* occurred in the Netherlands in 1988 (Vierbergen 1996). If *T. Palmi* established in the UK, glasshouse grown aubergines, chrysanthemums, cucumbers, Cyclamen, Ficus, orchids and sweet peppers would be principally at risk. Phytosanitary measures designed to inhibit the establishment of *T. Palmi* in Europe have been justified through pest risk analysis (MacLeod and Baker 1998). As a quarantine pest for the European Union (EU) (European Commission 2000) the UK National Plant Protection Organisation (NPPO) policy is to exclude *T. Palmi*, destroying interceptions and eradicating outbreaks if they occur. The first ever UK outbreak of *T. Palmi* was confirmed at a glasshouse site producing ornamental cut flowers in April 2000. An intensive treatment programme including soil, compost, foliar and space treatments was undertaken to eradicate the pest (MacLeod et al. 2004) and no movement of planting material from the site was allowed. After 15 months of intensive treatment and monitoring, eradication was declared in July 2001. To achieve eradication both the business at the outbreak site and the NPPO had to incur significant costs. An ex-ante BCA was therefore conducted to investigate whether such expenditure had been justified.

The insecticide application records before and during the eradication campaign were compared to determine the additional chemical-treatment costs. Together with the costs of additional hygiene measures, the cost of the outbreak to the producer was approximately £ 1,835 ha^{-1} month^{-1} (derived from MacLeod et al. 2004). This was more than six times the normal monthly cost for pest control at the site. Based on labour inputs, NPPO costs during the eradication campaign were approximately £ 123,000. In total the combined eradication costs to the NPPO and the grower were approximately £ 178,000.

Modelling the economic impacts of Thrips palmi spread from the outbreak site

A model was developed to assess the potential economic impact that could result from the spread of *T. Palmi* from the outbreak site had an eradication policy not been followed. The model considered the expansion of *T. Palmi* through protected horticulture in the UK over ten years at two rates. First, a rapid rate similar to the spread of *Frankliniella occidentalis*, a thrips species that previously spread through UK glasshouses over a three-year period from its first finding in June 1986 (Jones et al. 2005). Secondly, a slower rate, based on *T. Palmi* in Japan, where 62.5 % of the endangered area became occupied over ten years. Estimates of potential yield losses projected over 10 years were based on damage reports in the literature (Table 1).

Table 1. *Basic data for glasshouse crops at risk from Thrips palmi in the UK*

Crop	Area at risk (ha)	Crop value (£ '000)	Potential yield loss (%)	Value of potential losses (£ '000)	Ref. for yield loss estimate
Cucumbers	172	38,539	10	3,854	Kawai (1986)
Protected ornamentals	99[a]	14,705	1	147	See text
Sweet peppers	48	7,799	8	624	Nuessley and Nagata (1993)
Aubergines	11	2,548	15	382	Nagai (1991)
	330	63,591		5,007	

[a] Out of 990 ha protected ornamental production, 10 % was assumed to be *T. Palmi* hosts

There are no quantitative reports of *T. Palmi* damage to ornamental hosts. Given the polyphagy of *T. Palmi*, it has been estimated that about 10% of total ornamental production is susceptible (Mumford et al. 2000) and that in serious outbreak years feeding damage would cause a reduction of between 1 and 10 % in the value of affected hosts, due to losses in yield and/or quality and increased pest control costs (Kehlenbeck 1996). Accordingly, it was estimated that additional annual financial costs worth approximately 1 % of the value of protected ornamental production would be incurred within the infested area. The difficulty in estimating potential impacts was compounded by uncertainty in the level of damage caused by *T. Palmi* populations in each year. Estimates of severe damage, or high impact, in which populations caused maximum damage were taken from data in the existing literature and low or less severe impacts, where populations caused ten times less damage, as suggested by Kehlenbeck (1996), were selected to represent the range of possible impacts. It was anticipated that during the period of *T. Palmi* spread government and industry would undertake research to investigate ways to control and limit *T. Palmi* damage. Consequently costs of £ 50,000 per annum were allowed for. As a quarantine pest in four continents, exports of hosts liable to carry *T. Palmi* could be lost. Additional export certification could mitigate such losses. Such additional costs would be borne by government. Tables 2 and 3 show model outputs summarizing the annual area of glasshouse occupied from scenarios of fast spread (Table 2) and slow spread (Table 3) together with the present value of economic impacts caused by *T. Palmi* under scenarios of high and low impact. The HM-Treasury-recommended discount rate at the time of the *T. Palmi* outbreak was 6.5 % (HM Treasury 1997) and was used to determine present values.

Table 2. *Glasshouse area occupied and present value of projected economic impacts in a scenario of fast Thrips palmi spread with high or low impacts*

Year	Glasshouse area occupied (ha)	Present value of industry and government costs				Combined costs	
		High crop impact (£ '000)	Low crop impact (£ '000)	R & D (£ '000)	Additional certification (£ '000)	High (£ '000)	Low (£ '000)
1	2	3	0	47	21	71	68
2	82	111	11	45	20	175	75
3	248	316	32	42	18	377	92
4	330	397	40	40	17	454	97
5	330	374	37	37	16	428	91
6	330	353	35	35	16	404	86
7	330	333	33	33	15	381	81
8	330	314	31	31	14	359	77
9	330	296	30	30	13	339	72
10	330	279	28	28	12	320	68
						3,306	807

Table 3. Glasshouse area occupied and present value of projected economic impacts in a scenario of slow Thrips palmi spread with high or low impacts

Year	Glasshouse area occupied (ha)	Present value of industry and government costs				Combined costs	
		High crop impact (£ '000)	Low crop impact (£ '000)	R & D (£ '000)	Additional certification (£ '000)	High (£'000)	Low (£ '000)
1	2	3	0	47	21	71	68
2	26	35	4	45	20	99	68
3	49	62	6	42	18	123	67
4	69	83	8	40	17	140	65
5	92	104	10	37	16	158	64
6	116	124	12	35	16	175	63
7	139	140	14	33	15	188	62
8	158	150	15	31	14	195	60
9	181	163	16	30	13	205	59
10	204	173	17	28	12	213	57
						1,567	634

The financial benefits resulting from *T. Palmi* exclusion can be considered as the costs that are avoided if the UK had to 'live with' the pest. Comparing the range of the benefits from eradication with the costs involved in achieving eradication gives a series of benefit/cost ratios (Table 4).

Table 4. Benefit/cost ratios of Thrips palmi eradication

		Rate of spread	
		Fast	slow
Economic impact during *T. Palmi* spread	High	3,306 : 178 (19 : 1)	1,567 : 178 (9 : 1)
	Low	807 : 178 (5 : 1)	634 : 178 (4 : 1)

The benefit/cost ratios range from 4:1 to 19:1, depending upon the rate of spread and whether impacts are low or high and show that the policy of eradication was justified. The potential loss of exports is not included in the above analysis. Had such losses been included then the benefits of exclusion would be much higher (MacLeod et al. 2004).

EXAMPLE 2: THE ERADICATION OF *RALSTONIA SOLANACEARUM* FROM
THE RIVER TRENT BY REMOVAL OF HOST PLANTS FROM RIVERBANKS

Background

Potato brown rot is caused by the bacterium *Ralstonia solanacearum* race 3, biovar
2, and is a serious EU quarantine disease of potatoes that severely limits potato
production in temperate and tropical regions of the world (Smith et al. 1997). The
disease can be distributed via seed and ware potatoes, potato waste and, crucially for
the purposes of this analysis, via contaminated river water used for irrigation. In
accordance with EU Council Directive 98/57/EC, contaminated watercourses are
designated and, to inhibit spread of the bacterium, irrigation of potatoes using
contaminated water is prohibited.

Potatoes are the most extensively irrigated crop in the UK (MacKerron 1993).
Irrigation can have a beneficial impact on yields, especially for early potato
varieties. Irrigation also helps to control tuber quality, especially the incidence of
common scab, caused by the soil-borne bacterium *Streptomyces scabies*. A British
Potato Council analysis shows that, over four seasons, lack of irrigation led to a 40
% yield penalty and a 2-3-fold increase in common scab on packing varieties.
Because the effect of irrigation on yield and quality is so large, potato growers
always ensure that crops are irrigated whenever possible (Allen and Scott 1992).

Solanum dulcamara is a common native plant that can be found throughout the
UK. Where *S. dulcamara* grows on riverbanks and extends its roots into water
contaminated with *R. solanacearum*, it can become infected with *R. solanacearum*
allowing the bacterium to persist and provide a source of inoculum to further
contaminate the water. In 1998 the UK NPPO began a trial programme to remove *S.
dulcamara* from the River Great Ouse and the River Nene, which were designated
contaminated waterways. Annual costs for the removal of *S. dulcamara* averaged
£1,260 km^{-1} of river (2003 prices) (MacLeod 2004).

Removing Solanum dulcamara from banks of the River Trent

During a national survey of major waterways in 2003, the River Trent and canals
and tributaries linked to the Trent were found to be infected with *R. solanacearum*
(Defra 2003). At 274 km long, the Trent is one of the major rivers of England. Using
a geographic information system it was estimated that *S. dulcamara* would have to
be managed along 210 km of watercourses in the region. Local NPPO officers
responsible for the area estimated that 744 ha of potatoes would be affected by a ban
on using irrigation water from the River Trent and designated watercourses in the
region. In order to investigate whether the policy of enforcing an irrigation ban was
justified, an ex-ante BCA was conducted. Based on costs of previous work on the
rivers Great Ouse and Nene, the cost of removing *S. dulcamara* from the River Trent
was calculated to be approximately £ 265,000 per annum.

Potato growers who irrigate from the Trent obtain relatively high yields of
between 48 and 53 t ha^{-1}. Based on the gross margin budget for high-yielding
potatoes (Nix 2003), a grower with these yields may expect margins of between

£ 2,000 and £ 2,400 ha^{-1}. Over the 744 ha potentially affected by the irrigation ban, aggregate margins of between £ 1.5 million and £ 1.8 million would therefore be expected. Without irrigation, margins could fall by 15 to 40 % with losses of £ 0.9 million to £ 1.5 million. To avoid such losses, farmers could invest in pumps that can incorporate peroxygen chemical disinfectants that cleanse contaminated water by rapidly oxidizing organic matter killing *R. solanacearum*, thereby allowing crops to be irrigated. This would maintain potato yield and quality (BPC 2002). Using such disinfectants would increase variable costs by between £ 282 and £ 353 ha^{-1}, which amounts to approximately £ 232,000 to £ 266,000 across the 744 ha at risk. However, potatoes from fields irrigated with disinfected water would have to be tested for the presence of *R. solanacearum* in the first year before they were marketed. This would cost £ 115 per 25 tonnes (CSL 2003) and would probably be borne by the NPPO. Testing costs could amount to a one-off cost of between £ 164,000 and £ 181,000.

Benefits and costs of the phytosanitary campaign on the River Trent

Results from similar work on other rivers suggest that it takes at least four years to remove *S. dulcamara* and eradicate *R. solanacearum* from watercourses. The present value of industry and government costs during a four year programme on the River Trent is in the range of £ 2,058,000 to £ 2,205,000 (Table 5).

Table 5. *The present value of costs for a four year campaign on the River Trent*

Year	NPPO costs (£ '000)		Industry costs (£ '000)		
	Removal of *S. dulcamara*	Tuber testing	Irrigate with disinfectant	Discount factor*	PV sum of costs (£ '000)
1	265	164 to 181	232 to 266	1.000	661 to 712
2	265	0	232 to 266	0.966	477 to 510
3	265	0	232 to 266	0.943	467 to 499
4	265	0	232 to 266	0.902	447 to 478
					2,058 to 2,205

*The most recent recommended HM Treasury discount rate of 3.5% was used in this analysis. The rate differs from that used in example 1 since HM Treasury revised its recommendations in the light of existing and forecast changed economic circumstances (HM Treasury 2003)

Assuming success was achieved in the fourth year, then irrigation bans would be removed, growers' margins would return to existing levels and the extra costs of £ 232,000 to £ 266,000 incurred, due to disinfecting water, would be removed in perpetuity assuming that *R. solanacearum* did not return. The present value of these perpetuities would be from £ 5,777,000 to £ 6,623,000. Comparing the benefits of maintaining growers' gross margins in perpetuity after *R. solanacearum* has been eradicated, with costs of a four-year programme of *S. dulcamara* removal, the benefit/cost ratios vary from 2.6 : 1 to 3.2 : 1 (Table 6).

Table 6. *Benefit/cost ratios of removal of Solanum dulcamara from banks of the River Trent*

	From (£ '000)	To (£ '000)
From (£'000)	5,777 : 2,058 (2.8 : 1)	6,623 : 2,058 (3.2: 1)
To (£'000)	5,777 : 2,205 (2.6 : 1)	6,623 : 2,205 (3.0 : 1)

However, there is some uncertainty as to how long it may take to achieve eradication of *S. dulcamara* from the river so benefit/cost ratios were calculated for campaigns lasting 1 to 10 years. Figure 1 shows that the mean benefit/cost ratio improves if eradication can be achieved in under four years. Campaigns that last up to ten years approach the point where there is no net benefit in implementing the policy.

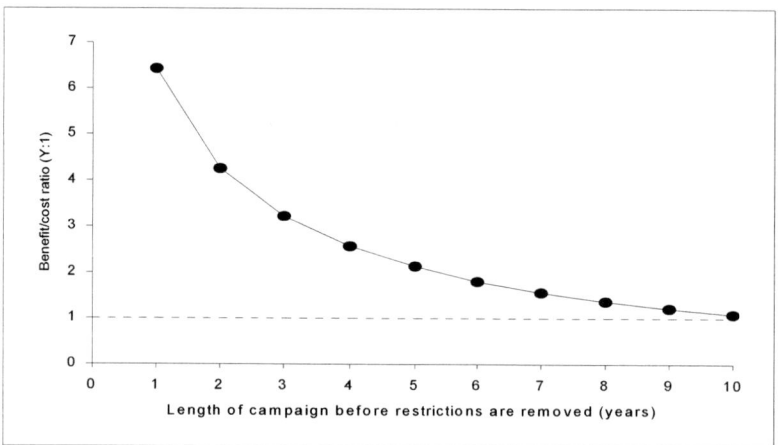

Figure 1. *Changes to the mean benefit/cost ratio through time*

The BCA study was used to inform policy with regard to the River Trent. Given the low benefit/cost ratio and the uncertainty in achieving eradication within four years, and bearing in mind the assumptions used in the analysis, a policy of removal of *S. dulcamara* from the banks of the River Trent was not followed.

EXAMPLE 3: BENEFIT/COST ANALYSIS OF EUROPEAN COMMUNITY MEASURES AGAINST THE WESTERN CORN ROOTWORM (*DIABROTICA VIRGIFERA VIRGIFERA*) IN THE UK

Background

Diabrotica virgifera virgifera, the western corn rootworm (WCR), is a univoltine oligophagous chrysomelid beetle from North America where it is one of the two most serious pests of continuous grain maize (Oerke et al. 1994). WCR was first detected in Europe in the former Yugoslavia in 1992. It is thought that WCR may have been introduced in 1990 by military air transport from North America (EPPO 1996). Since its arrival, WCR has spread annually with severe damage being reported for the first time in 1996. Due to the threat that this pest poses to EU member states the insect was added to the list of regulated pests in EU Plant Health legislation in January 1998 (European Commission 1998). Rotating maize with other crops can provide good control of WCR. Specific EU management measures designed to inhibit the spread of WCR include delay of harvest, use of insecticides and the restriction of growing maize within 1 km of an infested field for two years. Despite such measures WCR has spread within the EU and by June 2003 was detected in five of the then 15 EU countries (EPPO 2004) including the UK (Cannon et al. 2005). Following the finding, an ex-ante BCA was conducted to assess the impact of implementing the specific EU management measures designed to limit WCR spread.

The western corn rootworm in the UK

In the UK the vast majority of maize is grown for animal feed. MacLeod et al. (2005) developed two alternative scenarios and, using a stochastic Monte Carlo simulation model, annual costs associated with each scenario were estimated for a ten-year period. The first scenario estimated costs resulting from yield losses in continuous maize as a consequence of the NPPO not implementing EU measures. Annual rates of spread from the initial sites of infestation in SE England were selected from a triangular probability distribution, with parameters based on spread reported in the literature. By identifying the regions where maize is grown and overlaying it with climatic areas suitable for WCR development, the annual area of endangered maize could be determined (Baker et al. 2003). The model also used a triangular distribution of the minimum, most likely and maximum annual maize area suitable for WCR development. The model combined the annual area suitable for WCR establishment with the annual rate of spread and provided output in the form of maize area occupied by WCR each year and projected forward ten years. From 10,000 model iterations, the mean annual area occupied was used to calculate potential future losses in yield in unrotated maize. Evidence from European countries suggests that there is a time lag of approximately five years between the first finding of WCR and reports of economic damage in continuous maize (EPPO 2003). Around 20 % of maize in UK is grown continuously and hence is most at risk from WCR. Yield losses are predicted to be between 10 and 30 %. The second scenario estimated the costs to maize growers of implementing the EU measures. The annual area suitable for WCR establishment was randomized but used a slower rate of WCR spread. Under EU regulations, once a maize field is found to be infested with WCR, EU measures, for example, rotation, should be implemented in the field and all other surrounding maize fields within a 1-km Focus Zone. Measures are also required in an outer Safety Zone, extending from 1 to 6 km from the infested field. Cannon et al. (2005) describe the measures applied in the Focus and Safety Zones in the UK during 2003 and 2004.

Scenario 1: Costs of not implementing EU measures
Without implementing EU measures, the first scenario showed that, on average, WCR would continue to spread for three years before stabilizing to occupy just over 39,000 ha of maize each year. Yield losses would be seen in continuous maize after five years. The present value of aggregate losses after ten years ranges from £ 1.9 million to £ 2.3 million (MacLeod et al. 2005). However, the vast majority of maize in the UK can be rotated and growers would not suffer significant additional costs from implementing rotation if WCR became established in the UK.

Scenario 2: Costs of implementing EU measures
By implementing the EU measures, maize growers growing continuous maize with severe constraints to change may incur additional costs averaging between £ 182 ha-1 for fields in the Safety Zone and £ 243 ha-1 for fields in the Focus Zone. Over a ten-year period of WCR spread, during which almost 7,200 ha of maize with severe constraints to rotation become infested, the impact of implementing EU measures on maize growers would have a present value of approximately £ 14.7 million (MacLeod et al. 2005). Under the statutory campaign, no yield losses would be incurred since populations of WCR are prevented from reaching damaging levels. Nationally, approximately 2 % of Inspectorate time was spent on WCR activities during 2004-2005. By apportioning the NPPO financial budget to areas of Inspectorate activity, it is estimated that implementation of the existing policy costs around £ 228,000 per year.

Benefit/cost analysis for WCR
Summing industry and government costs of implementing EU measures for the next 10 years, and comparing them with expected losses as a result of living with WCR, provides benefit/cost ratios of between 1 : 8 and 1 : 7. The stochastic model used to estimate the benefits and costs of implementing EU WCR control measures in the UK shows that strict implementation of the measures does not appear to be economically justified over the next ten years. Management measures, especially the prohibition of growing maize in demarcated zones, can impose substantial costs on maize growers who have severe constraints to change. In contrast, with no statutory measures in place, yield losses caused by WCR in continuous maize are likely to be significantly lower than the cost of measures resulting from forced rotation. Costs resulting from a forced change in rotation are potentially substantial for some

growers and, whilst it is acknowledged that assessing the cost of a change in rotation is difficult (Baufeld 2003) and thus not included amongst the costs of impacts considered by the EU *Diabrotica* project by Vidal (2003), not including such costs can seriously underestimate the impact of management measures on maize growers. The present UK policy with regard to WCR is to adopt a *light* approach (Cannon et al. 2005) that balances the EU requirements with a pragmatic approach to pest management.

DISCUSSION

The examples given show how BCA can be a useful tool providing information for phytosanitary policy decision making. BCA has a number of strengths that make it a useful technique. Firstly it requires scenarios to be formulated, providing a structured framework within which costs and benefits are identified and quantified. Taking such an approach, the *T. Palmi* BCA reinforced the justification for implementing an eradication campaign. The analysis could inform future policy decisions if an outbreak of *T. Palmi* were to occur in the UK again. BCA provides a useful mechanism for dealing with uncertainty and complexity, but it does not make difficult problems simple. There are always uncertainties concerning the size of costs and benefits and the likelihood that such costs and benefits will occur. Whilst a BCA justified removal of *S. dulcamara* from the Trent, uncertainties about implementing such a policy were so significant that removal action was not taken. Similarly uncertainties about the future climate and its influence on WCR (Baker et al. 2003) led to the light management of WCR despite the BCA showing that measures were not currently economically justified.

When examining plant pests, BCA studies will often assess the primary and clearest costs and benefits, such as loss in yield or quality and use of additional pesticides. These directly affect producers whose crops are at risk. However, this is a very narrow view of the economic impacts caused by quarantine pests (Bigsby 2001). Such analysis does not take into account other potential indirect costs and benefits, secondary and tertiary effects, which may result (FAO 2001). However, measuring indirect effects on non-market goods is a difficult process and is a general weak point in many BCA studies, including the examples provided. BCA studies that evaluate decisions that may have environmental consequences often encounter such difficulties. The weakness is partly due to the complex nature of ecosystems and the difficulty in forecasting the effects of decisions regarding one part of an ecosystem and its bearing on another part of the ecosystem (Hanley 1990). Without understanding dependencies and relationships between constituents of the ecosystem, there is a considerable challenge to design economic tools that can fully quantify all impacts that stem from particular decisions. Thus there is scope for research to improve methods for measuring and incorporating secondary economic effects into BCA. By assessing the impact on producers and other sectors of the economy and on the environment, decisions would then be made for the benefit of society as a whole, not just for agricultural or horticultural producers. Unless all costs and benefits are properly valued, and considered by policy decision makers, society will not receive the optimum output from the resources available and economic inefficiency will result. Nevertheless, an alternative, pragmatic approach can be adopted in economies such as the UK, where crops provide only 0.46 % of the GDP and horticulture contributes approximately 0.25 % to GDP (*Britain 1999: the official yearbook of the United Kingdom* 1999). In such an economy it is unlikely that additional costs from specific pest impacts will feed back into the national economy to such an extent that such elaborate analyses are justified. This is not to say that within any particular sector, pest impacts would be negligible. For example, forecasts of the potential economic impact of quarantine pests can be substantial, e.g., the tobacco whitefly, *Bemisia tabaci*, would have great impact on producer costs within the UK tomato industry (Morgan and MacLeod 1996) whilst an Asian longhorn beetle, *Anoplophora glabripennis*, can cause severe impacts on hardwood urban and amenity trees with significant recreational value (MacLeod et al. 2002).

BCA is not a panacea and in addition to the weaknesses identified above, it must be recognized that in developing scenarios, it is difficult to forecast producers' behavioural responses and what action they will take when adapting to particular pests. With a large amount of uncertainty there may be some considerable variation in any benefit/cost ratio making it difficult to interpret results. Where this is the case further research may be necessary to reduce uncertainty. However, this could add significantly to the time taken to conduct BCA. Finally it is recognized that BCA often compares between choices of control or no control. As such BCA does not provide information about the marginal effect of control, i.e., what is the effect of one more or one less unit of control effort? Thus BCA does not determine what the appropriate level of control should be (FAO 2001).

There have been calls for BCA to be adopted more widely by governments when considering quarantine regulations (Robertson 2001), but BCA can never be the only basis upon which policy decisions are made, especially in relation to phytosanitary matters, due to the complex nature of ecosystems. The complex inter-relationships between species and their interaction with a changing environment make it difficult to predict the biological consequences of pest introductions. Determining their potential economic impacts is equally complicated. However, this is the challenge that faces those that work in the phytosanitary arena and it provides opportunities for economists and biologists to collaborate to overcome such difficulties.

REFERENCES

Allen, E.J. and Scott, R.K., 1992. Principles of agronomy and their application in the potato industry. *In:* Harris, P.M. ed. *The potato crop: the scientific basis for improvement.* Chapman and Hall, London, 816-881.

Baker, R.H.A., Cannon, R.J.C. and MacLeod, A., 2003. Predicting the potential distribution of alien pests in the UK under global climate change: *Diabrotica virgifera virgifera. In: The BCPC international congress: crop science and technology: proceedings of an international congress held at the SECC, Glasgow, Scotland, UK, 10-12 November 2003.* British Crop Protection Council, Alton, 1201-1208.

Baufeld, P., 2003. Technical dossier on pecuniary losses and costs of containment measures in high-risk areas. *In:* Vidal, S. ed. *Threat to European maize production by the invasive quarantine pest Western Corn Rootworm (Diabrotica virgifera virgifera): a new sustainable crop management approach.* EU Research Report no. QLRT-1999-01110.

Bigsby, H.R., 2001. The 'appropriate level of protection': a New Zealand perspective. *In:* Anderson, K., McRae, C. and Wilson, D. eds. *The economics of quarantine and the SPS agreement.* Centre for International Economic Studies, Adelaide, 141-163.

BPC, 2002. *Monitoring and control of the potato brown rot bacterium in irrigation water.* British Potato Council, Oxford. British Potato Council Factsheet.

Britain 1999: the official yearbook of the United Kingdom, 1999. Office for National Statistics, London.

Cannon, R.J.C., Matthews, L., Cheek, S., et al., 2005. Surveying and monitoring western corn rootworm (*Diabrotica virgifera virgifera*) in England and Wales. *In:* Alford, D.V. and Backhaus, G.F. eds. *Plant protection and plant health in Europe: introduction and spread of invasive species: symposium held at Humboldt University, Berlin, Germany, 9-11 June 2005.* British Crop Protection Council., Alton, 155-160.

CSL, 2003. *Plant pest and disease identification price guide 2003/04.* Central Science Laboratory, Sand Hutton.

Defra, 2003. *Potato brown rot: 2003 monitoring programme: final results.* Department for Environment, Food and Rural Affairs. [http://www.defra.gov.uk/planth/pbr2.htm]

EPPO, 1996. Situation of *Diabrotica virgifera* in Serbia (YU). From: International Workshop "Western Corn Rootworm in Europe 95", Gödöllö (HU), 1995-11-08. *EPPO Reporting Service,* 1, 96/006. [http://archives.eppo.org/EPPOReporting/1996/Rse-9601.doc]

EPPO, 2003. Situation of *Diabrotica virgifera virgifera* in the EPPO region. From: Papers presented at the 7th Meeting of the EPPO ad hoc Panel on *D. virgifera* held jointly with the 9th International IWGO Workshop on *D. virgifera* in Belgrade, 2002-11-03/05. *EPPO Reporting Service,* 1, 2003/001. [http://archives.eppo.org/EPPOReporting/2003/Rse-0301.doc]

EPPO, 2004. *EPPO Plant quarantine information retrieval system, v4.3.* EPPO, Paris. [http://www.eppo.org/PUBLICATIONS/pqr/pqr.htm]

European Commission, 1998. EC Directive 98/1/EC of 8 January 1998 amending certain Annexes to Council Directive 77/93/EEC on protective measures against the introduction into the Community of organisms harmful to plants or plant products and against their spread within the Community. *Official Journal of the European Communities,* L 15 (21.1.1998), 26-33. [http://eur-lex.europa.eu/LexUriServ/site/en/oj/1998/l_015/l_01519980121en00260033.pdf]

European Commission, 2000. EC Council Directive 2000/29/EC of 8 May 2000 on Protective measures against the introduction into the Community of organisms harmful to plants or plant products and against their spread within the Community. *Official Journal of the European Communities,* L 169/1 (10.7.2000), 1-112 and subsequent amendments. [http://europa.eu.int/eur-lex/pri/en/oj/dat/2000/l_169/l_16920000710en00010112.pdf]

FAO, 2001. Economic impacts of transboundary plant pests and animal diseases. *In: The state of food and agriculture 2001.* FAO, Rome. [http://www.fao.org/docrep/003/x9800e/x9800e14.htm]

FAO, 2003. *Pest risk analysis for quarantine pests, including analysis of environmental risks and living modified organisms.* Secretariat of the International Plant Protection Convention FAO, Rome. ISPM no. 11. [http://www.fao.org/DOCREP/006/Y4837E/Y4837E00.HTM]

Hanley, N., 1990. Are there environmental limits to cost benefit analysis? *In:* Van den Noort, P.C. ed. *Costs and benefits of agricultural policies and projects: proceedings of the 22nd symposium of the European Association of Agricultural Economists EAAE, October 12th-14th, 1989, Amsterdam, Netherlands.* Wissenschaftsverlag Vauk, Kiel, 202-212.

HM Treasury, 1997. *Appraisal and evaluation in central government: Treasury guidance.* HMSO, London.

HM Treasury, 2003. *The green book: appraisal and evaluation in central government.* The Stationery Office Books, London. [http://greenbook.treasury.gov.uk/]

Jenkins, P.T., 1999. Trade and exotic species introductions. *In:* Sandlund, O.T., Schei, P.J. and Viken, A. eds. *Invasive species and biodiversity management: papers presented at the Norway/United Nations UN conference on alien species, held in Trondheim, Trondheim, Norway, 1-5 July 1996.* Kluwer Academic Publishers, Dordrecht, 229-235. Population and Community Biology Series no. 24.

Jones, G.D., Perrings, C.A. and MacLeod, A., 2005. The spread of *Frankliniella occidentalis* through the UK protected horticulture sector:scales and processes. *In:* Alford, D.V. and Backhaus, G.F. eds. *Plant protection and plant health in Europe: introduction and spread of invasive species, held at Humboldt University, Berlin, Germany, 9-11 June 2005.* British Crop Protection Council., Alton, 13-18.

Kawai, A., 1986. Studies on population ecology of *Thrips palmi* Karny. X. Differences in population growth on various crops. *Japanese Journal of Applied Entomology and Zoology*, 30 (1), 7-11.

Kawai, A., 1990. Control of *Thrips palmi* in Japan. *Japan Agricultural Research Quarterly*, 24 (1), 43-48.

Kehlenbeck, H., 1996. *Nutzen-Kosten-Untersuchung "Auswirkungen der EG-Binnenmarktregelungen im Bereich der Pflanzengesundheit"*. Bonn, Köllen. Schriftenreihe des Bundesministers für Ernährung, Landwirtschaft und Forsten. Angewandte Wissenschaft no. 456.

Krystynak, R., 1991. The economic importance of plant protection programs in Canada. *Canadian Farm Economics*, 23 (1), 41-47.

Levine, J.M. and D'Antonio, C.M., 2003. Forecasting biological invasions with increasing international trade. *Conservation Biology*, 17 (1), 322-326.

MacKerron, D.K.L., 1993. The benefits to be gained from irrigation. *In:* Bailey, R.J. ed. *Irrigating potatoes.* Cranfield Press, Cranfield, 1-11. UK Irrigation Association Technical Monograph no. 4.

MacLeod, A., 2004. *Cost-benefit analysis of a campaign to remove Solanum dulcamara and eradicate Ralstonia solanacearum from the River Trent, and to end the current campaign on the River Witham/Fossdyke Navigation canal.* Unpublished Report to Defra Plant Health Division.

MacLeod, A. and Baker, R.H.A., 1998. Pest risk analysis to support and strengthen legislative control of a quarantine thrips: the case of *Thrips palmi. In: The BCPC international congress: crop science and technology: proceedings of an international conference, Brighton, UK, 16-19 November 1998. Volume 1.* British Crop Protection Council, Farnham, 199-204.

MacLeod, A., Baker, R.H.A. and Cannon, R.J.C., 2005. Costs and benefits of European Community (EC) measures against invasive alien species: current and future impacts of *Diabrotica virgifera virgifera* in England & Wales. *In:* Alford, D. and Backhaus, G. eds. *Introduction and spread of invasive species:symposium, 9-11 June 2005, Berlin, Germany.* British Crop Protection Enterprises. BCPC Symposium Proceedings no. 81.

MacLeod, A., Evans, H.F. and Baker, R.H.A., 2002. An analysis of pest risk from an Asian longhorn beetle (*Anoplophora glabripennis*) to hardwood trees in the European community. *Crop Protection*, 21 (8), 635-645.

MacLeod, A., Head, J. and Gaunt, A., 2004. An assessment of the potential economic impact of *Thrips palmi* on horticulture in England and the significance of a successful eradication campaign. *Crop Protection*, 23 (7), 601-610.

Morgan, D. and MacLeod, A., 1996. Assessing the economic threat of *Bemisia tabaci* and tomato yellow leaf curl virus to the tomato industry in England and Wales. *In: The BCPC international congress: crop science and technology: proceedings of an international conference, Brighton, UK, 18-21 November 1996. Volume 3.* British Crop Protection Council, Farnham, 1077-1082.

Mumford, J., Temple, M., Quinlan, M., et al., 2000. *Economic policy evaluation of MAFF's plant health programme: report to Ministry of Agriculture Fisheries and Food.* ADAS Consulting, London. [http://statistics.defra.gov.uk/esg/evaluation/planth/default.asp]

Nagai, K., 1991. Integrated control programs for *Thrips palmi* on eggplants (*Solanum melongena* L.) in an open field. *Japanese Journal of Applied Entomology and Zoology*, 35 (4), 283-289.

Nix, J., 2003. *Farm management pocketbook.* 34th edn. Wye College, London.

Nuessly, G.S. and Nagata, R.T., 1993. Pepper varietal response to thrips feeding. *In:* Parker, B.L., Skinner, M. and Lewis, T. eds. *Thrips biology and management: proceedings of the 1993 international conference on Thysanoptera.* Plenum Publishing, London, 115-118.

Oerke, E.C., Dehne, H.W., Schönbeck, F., et al., 1994. *Crop production and crop protection: estimated losses in major food and cash crops.* Elsevier, Amsterdam.

Pemberton, A.W., 1988. Quarantine: the use of cost:benefit analysis in the development of MAFF plant health policy. *In:* Clifford, B.C. and Lester, E. eds. *Control of plant diseases: costs and benefits.* Blackwell Scientific Publications, Oxford, 195-202.

Rautapaa, J., 1984. Costs and benefits of quarantine measures against Liriomyza trifolii in Finland. *EPPO Bulletin*, 14 (3), 343-347.

Robertson, D., 2001. Summing up. *In:* Anderson, K., MacRae, C. and Wilson, D. eds. *The economics of quarantine and the SPS agreement.* Centre for International Economic Studies, Adelaide, 387-395.

Smith, I.M., MacNamara, D.G., Scott, P.R., et al. (eds.), 1997. *Quarantine pests for Europe: data sheets on quarantine pests for the European Union and for the European and Mediterranean Plant Protection Organization.* 2nd edn. CAB International, Wallingford.

Sumner, D.A., 2003. Exotic pests and public policy for biosecurity: an introduction and overview. *In:* Sumner, D.A. ed. *Exotic pests and diseases: biology, economics and public policy for biosecurity.* Iowa State Press, Ames, 3-6.

Vidal, S., 2003. *Threat to European maize production by the invasive quarantine pest Western Corn Rootworm (Diabrotica virgifera virgifera): a new sustainable crop management approach.* EU Research Report no. QLRT-1999-01110.

Vierbergen, G., 1996. After introduction of *Frankliniella occidentalis* in Europe: prevention of establishment of *Thrips palmi* (Thysanoptera: Thripidae). *Acta Phytopathologica et Entomologica Hungarica,* 31 (3/4), 267-274.

ECONOMIC
AND BIOPHYSICAL
ASPECTS OF PLANT
HEALTH POLICIES

CHAPTER 12

MODEL FRAMEWORKS FOR STRATEGIC ECONOMIC MANAGEMENT OF INVASIVE SPECIES

JOHN D. MUMFORD

*Centre for Environmental Policy, Imperial College London,
Silwood Park, Ascot SL5 7PY, United Kingdom*

Abstract. To allocate biosecurity resources efficiently and effectively it is necessary to be able to systematically estimate and describe risks from a wide range of threats and mitigation measures. A common framework for conducting risk assessments is an essential tool for setting national priorities and for making decisions that will justify actions to international trading partners. Two systems, one quantitative and one that combines qualitative and quantitative elements, are presented as examples of such a systematic approach.

Keywords: biosecurity; invasive pests; risk assessment; risk framework; risk profile

INTRODUCTION

Much of the literature on economic aspects of invasive species has focussed on estimates of historical losses, for example the Office of Technology Assessment (OTA 1993) report to the US Congress. Historical losses are useful evidence of the potential and likely scale of analogous new invasions and provide a justification for the continued allocation of resources to quarantine actions against such pests. Quarantine authorities are faced with the constant problem of deciding how to balance their limited efforts to prevent, intercept, detect and/or eradicate specific threats from amongst the thousands of potential pest species that may enter a country through a wide range of pathways. The World Trade Organisation Agreement on the Application of Sanitary and Phytosanitary Measures requires signatories to base their actions on sound scientific evidence and a consistent level of acceptable risk, based on appropriate international standards. The International Plant Protection Convention, for example, offers standards on Pest Risk Analysis methods for quarantine pests (FAO 2003). Stohlgren and Schnase (2006) describe the iterative steps involved in establishing the level of risk from individual species and pathways.

The strategic management of invasive species requires a consistent framework for economic assessment across the wide range of potential species involved.

A.G.J.M. Oude Lansink (ed.), New Approaches to the Economics of Plant Health, 181–190.
© 2007 *Springer*.

Without a common method to predict the impacts it would not be possible to establish priorities on risk reduction or make assessments of the share of responsibilities that might be apportioned to the various participants in the system. This applies prior to introductions (Mumford 2002) as well as for decisions on eradication or suppression subsequent to an outbreak or long-term establishment of an exotic pest (Mumford 2005). Decisions concerning invasive species that affect natural environments in particular need to be included in a consistent framework to ensure that environmental impacts are not ignored, although this does not guarantee the priority for such species will be high (Mumford 2001).

Policy makers are faced with major issues at several levels regarding invasive species. The National Audit Office (2003) described the various policy needs for England and Wales, and the needs for many other countries would be similar. There are strategic decisions about the key invasive pest species on which to focus actions, for which preventative actions and pre-planned emergency measures for outbreaks can be budgeted, based on expected probabilities of detection. However, other species will also be detected in the course of routine inspection and surveillance and tactical decisions must be made rapidly on what actions to take and how to fund prevention, containment or control. All of these decisions require common frameworks to ensure that appropriate responses are justified and that the level of risk is consistent.

The following sections illustrate two mechanisms for establishing consistent frameworks for invasive-species assessment. The first involves a stochastic modelling process with generic variables describing invasion and impact, and the second is a more subjective classification scheme for systematically describing risk and impact for either planned or accidental introductions.

GENERAL QUANTITATIVE FRAMEWORK FOR INVASIVE PEST PRIORITIES

The need for a consistent framework for assessing invasive species has been recognized by the Department for Environment, Food and Rural Affairs (DEFRA) in the United Kingdom, in part as a way of managing the changing responses to a constantly evolving problem of invasion (Waage et al. 2004). New challenges arise from invasive species through new pathways, increased volume of trade or climate change, public attitudes put different relative values on environmental conservation, agricultural production or animal welfare, and technological developments create new pest prevention, detection and management opportunities, as well as potential new pathways (Waage et al. 2005). Despite such constantly changing circumstances predictions of impact must still be made, and they must include estimates of the uncertainties that come with that change.

Invasions follow a generic pattern of entry, establishment, spread and growth leading to impact over some proportion of the resource affected. Most do not succeed at some stage in this process (Williamson 1996) and the probability of failure should be accounted for in estimates of impact. A stochastic approach is needed to describe the range of outcomes that are possible over time as repeated invasion opportunities occur.

Waage et al. (2004) presented a demonstration of how a single generic model could include sufficient flexibility and detail to provide a useful estimate of impacts from a wide range of potential invasive species in the United Kingdom (Figure 1). The model consists of ecological modules that lead to estimates of the extent of invasion using common parameters related to success of entry, establishment, spread and growth of populations, taking into account potential control actions. These are coupled to economic modules that put values on the damage and control efforts associated with the added impact of a new pest over a 20-30-year time horizon.

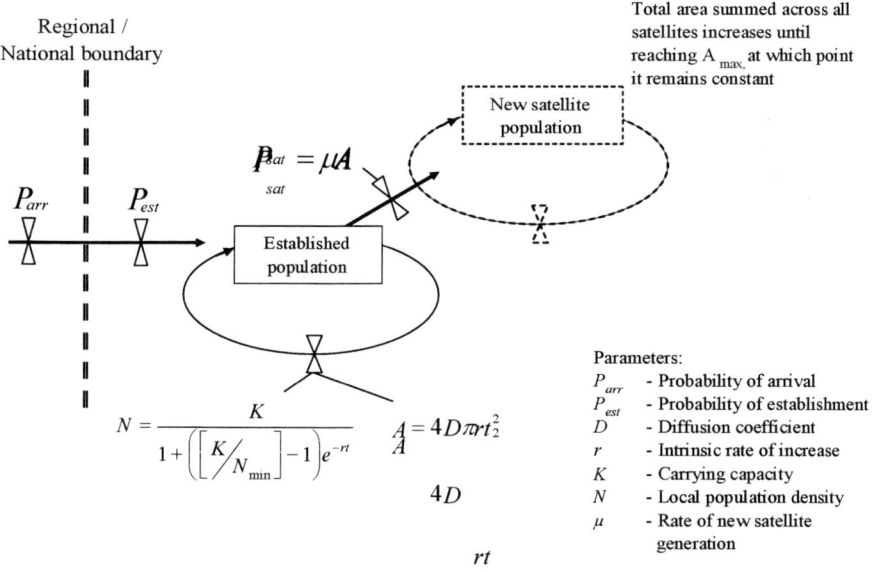

Figure 1. *Schematic representation of a generic entry, establishment, spread and growth model (from Waage et al. 2004)*

Figure 2 demonstrates the typical curves of impact over time for three categories of invasive pests, those that attack crops, livestock and natural environments. The vertical axis represents the total value of a sector that could be affected by an invasive pest, with 100 % being its pre-invasion value, with a subsequent decline in value related to the proportion of the resource that is affected by the new pest (shown on the horizontal axis). Crop pests cause damage that is more or less linearly related to the extent of the crop affected (unless there are significant effects on export potential, which can occur with some major plant quarantine pests). New livestock pests are much more likely to have an impact on export trade or travel, as occurred with the Foot and Mouth epidemic in the United Kingdom in 2001, causing significant immediate economic impact to the industries concerned even with only a limited presence. Pests in the natural environment often cause very little loss of value while they increase, because most of the quality of the environment is retained despite their presence. Eventually, however, as a much larger proportion of the environment is affected its overall value falls sharply as people begin to realize the rarity of the remaining unaffected portion. In each case, the ultimate measure that is used is the proportion of the resource and its value, which allows a common scaling from the model.

The long time delay affecting the impact from pests in the natural environment poses a serious problem in placing priority on such pests, because of the compounding effect of discounting (Mumford 2001). At the extreme, an invasive species may not even be noticed for some time in the natural environment. For example, the modal time lag from first introduction to discovery in the wild for new plant species in the United Kingdom is approximately 100 years (Preston et al. 2002). By contrast, the arrival of a notifiable human or animal disease should be recorded within days. Inevitably, pests of livestock are likely to have higher priority than plants because their impact is more likely to be immediate, and there is an important element of concern for animal welfare in the public, who are now also concerned about the possible crossover of animal diseases to humans (for example, avian influenza). This is reflected in the relative spending on animal and plant quarantine in the United Kingdom, where 90% of the funding supported animal health in 2000, even before the Foot and Mouth epidemic (Mumford 2002).

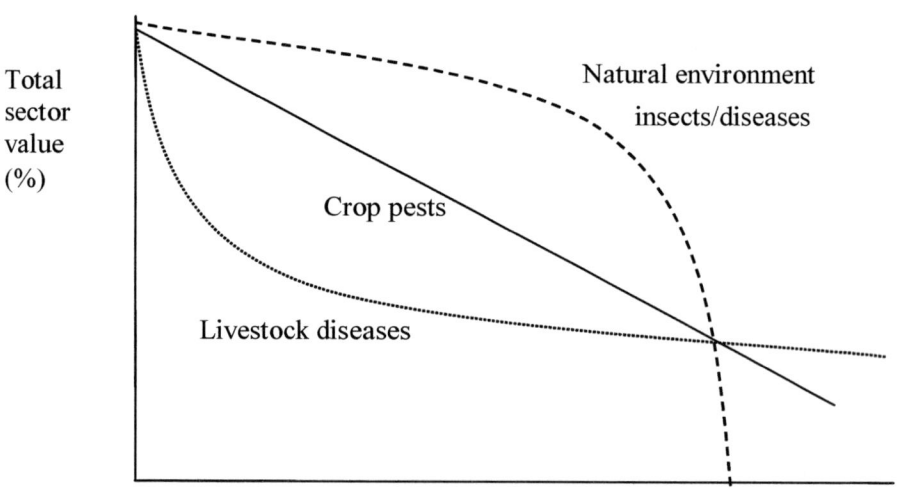

Figure 2. Three general curves depicting the economic relationship between the proportion of resource affected by an invasive pest and the impact on the total value of the sector affected (from Waage et al. 2004)

This approach allows systematic sensitivity analysis of parameter values to determine the effect of their contribution to the overall uncertainty in the estimation of impact. This could be used as the basis for additional research or subjective enquiry to narrow the uncertainty range for particularly variable parameters. Figure 3 illustrates an estimated risk distribution for annual Newcastle Disease loss and control costs in the United Kingdom for a 20-year time horizon. The model in this case is particularly sensitive to estimates of the probability of disease entry and establishment and the proportion of export revenue lost. For crop pests, spread and yield loss estimates are likely to be critical sensitivity parameters.

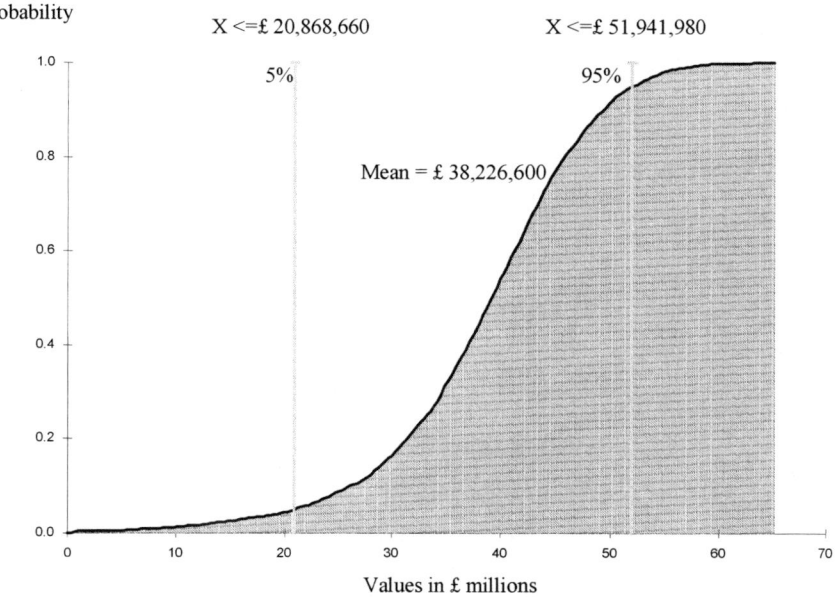

Figure 3. *Cumulative distribution of the level of expected annual damage over 20 years for Newcastle Disease, a disease of poultry (from Waage et al. 2004)*

A shortcoming with this approach is that the generic format limits inclusion of some detail that might be helpful in specific cases and requires parameter data to be estimated in a form that is not always clearly applicable to a particular species, given the diversity of ways in which pests enter, establish, spread and grow. For instance, the dispersal of plants by seeds, rhizome and transplanting involves a complex of parameter values for spread, while aquatic vertebrates spread linearly in rivers. However, the opportunity to make direct comparisons across a broad range of taxa and resources compensates for the restrictions and simplification imposed by the model structure. This approach is intended for general screening of priorities, which might be followed by more detailed, case-specific analyses of high-priority invasive species.

SUBJECTIVE RISK FRAMEWORKS FOR BOTH BENEFICIAL
AND HARMFUL NON-NATIVE INTRODUCTIONS

An attempt has been made by a multidisciplinary group in the United Kingdom to create a common system for assessing the impacts of non-native species that could enter the country, either by accident or design (Defra 2005). Previously, independent assessments using different criteria and scales have been applied to the various taxonomic groups by different responsible technical centres within DEFRA. The proposed new scheme consists of a standard set of questions related to the entry, establishment, spread and impacts of a new organism, which is designed to cover the full range of taxa that could enter, from pathogens to vertebrates. The module on economic impact includes questions to establish the magnitude (Table 1) and likelihood (Table 2) of introductions on common scales that can be combined to form an acceptability matrix (Table 3).

Table 1. *Magnitude values for risks, using four subjectively equivalent dimensions (from Defra 2005; and modified from Standards Australia 2004)*

Scale and score	Monetary loss and response costs	Health impact	Environment impact	Social impact
Minimal 1	Up to £ 10k /yr	Local, mild, short-term, reversible effects to individuals	Local, short-term population loss, no significant ecosystem effect	No social disruption
Minor 2	£ 10k - £ 100k /yr	Mild short-term reversible effects to identifiable groups, localized	Some ecosystem impact, reversible changes, localized	Significant concern expressed at local level
Moderate 3	£ 100k - £ 1m /yr	Minor irreversible effects and/or larger numbers covered by reversible effects, localized	Measurable long-term damage to populations and ecosystem, but little spread, no extinction	Temporary changes to normal activities at local level
Major 4	£ 1m - £ 10m /yr	Significant irreversible effects locally or reversible effects over large area	Long-term irreversible ecosystem change, spreading beyond local area	Some permanent change of activity locally, concern expressed over wider area
Massive 5	£10m+ /yr	Widespread, severe, long-term, irreversible health effects	Widespread, long-term population loss or extinction, affecting several species with serious ecosystem effects	Long-term social change, significant loss of employment, migration from affected area

Many pest risk assessments must be subjective because of the lack of verified data relevant to the specific issues of introduction and damage in a new environment. The framework shown in Table 1 is an attempt to provide a set of definitions over a range of independent dimensions that might be appropriate to potential invasive species. This system is based on the Australia/New Zealand Risk Management Standard (AS/NZS 4360 Risk Management), but with some modification of the monetary values, and of the wording in the other three dimensions. A logarithmic five-point scale of magnitude of risks can be applied, which allows an approximate translation of impacts to a monetary scale. The five-point range of orders of magnitude covers the main range in which there is a relatively routine decision problem (tens of thousands of £ to tens of millions of £). Where potential impacts are significantly greater than this (as they would be with Foot and Mouth Disease, for example) it is not a routine decision.

Table 2. *Likelihood of impacts with descriptions and frequencies (from Defra 2005; and modified from Standards Australia 2004)*

Likelihood and score	Description	Frequency
Very unlikely 1	This sort of event is theoretically possible, but is never known to have occurred and is not expected to occur	1 in 10,000 years
Unlikely 2	This sort of event has not occurred anywhere in living memory	1 in 1,000 years
Possible 3	This sort of event has occurred somewhere at least once in recent years, but not locally	1 in 100 years
Likely 4	This sort of event has happened on several occasions elsewhere, or on at least one occasion locally in recent years	1 in 10 years
Very likely 5	This sort of event happens continually and would be expected to occur	Once a year

If a risk is a single one-off loss, or a series of specific incidents, for instance outbreaks of a disease which are quickly eradicated, it should be converted into an annualized average present value using the discount rate over a predetermined time horizon. The time periods and the discount rates selected can have a major effect on the estimated annual loss for single-point events.

Generally a new organism or pest invasion would be expected to cause a continuing loss, and that it could increase in impact over time, as was indicated in Figure 2. If the magnitude is expected to grow then an average annual value based on a net present value of the expected flow of loss/cost could be used as the base value to determine average annual loss over the proposed time period.

The likelihood values (Table 2) are also on a log scale of frequency and are scored on a five-point scale. This system is also based on the Australia/New Zealand Risk Management Standard with modified wording of definitions, and shifting of the frequencies related to some descriptions to make them less frequent. The Standard uses seven categories, including 1/3-year and 1/30-year frequencies, which would be approximate intermediate values on a log scale between the three relatively frequent categories. Scale scores for magnitude and likelihood can be added to give an overall value of risk, because both are on log scales. Where more specific estimates are available for loss or likelihood, fractional scores could be used to make calculations more precise.

Uncertainty can be expressed by assigning probability values to the likelihood and magnitude scales. The acceptability of risk can be described, as shown in Table 3. 'Negligible', 'Justifiable' and 'Unacceptable' risk would be judged against the benefits or costs of prevention and should be defined in a way that can be applied to any particular taxonomic example.

The uncertainty values for the individual dimensions of likelihood and magnitude can be used to express the uncertainty of the combined risk levels graphically, as seen in three-dimensional form in Figure 4. This shows the extent of the uncertainty in each dimension, while focussing on the most likely outcome expected.

Table 3. *Risk acceptability values, with uncertainty values for the two axes*

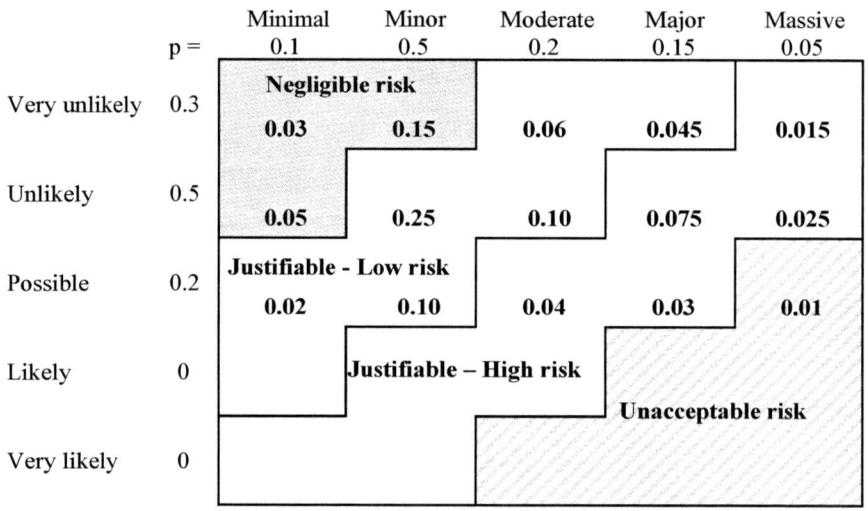

	p =	Minimal 0.1	Minor 0.5	Moderate 0.2	Major 0.15	Massive 0.05
Very unlikely	0.3	**Negligible risk** **0.03**	**0.15**	**0.06**	**0.045**	**0.015**
Unlikely	0.5	**0.05**	**0.25**	**0.10**	**0.075**	**0.025**
Possible	0.2	**Justifiable - Low risk** **0.02**	**0.10**	**0.04**	**0.03**	**0.01**
Likely	0	**Justifiable – High risk**				
Very likely	0			**Unacceptable risk**		

Cumulative probability of risk exceeding nominal value

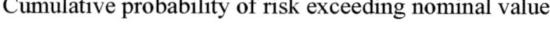
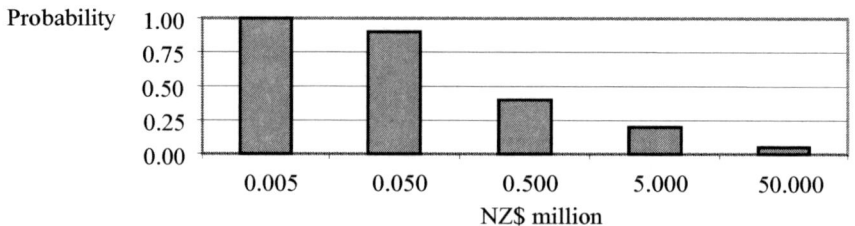

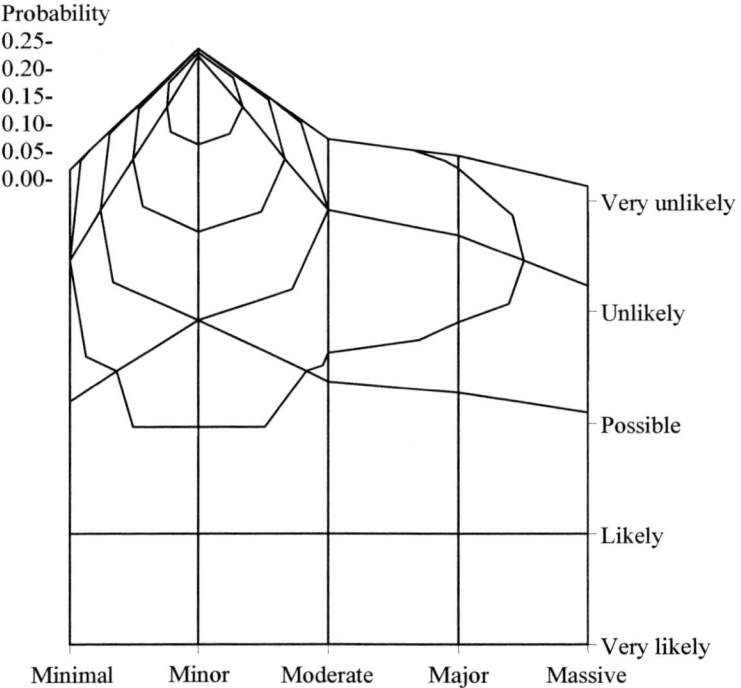

Figure 4. *An example risk profile, showing the focus of risk from table 3*

CONCLUSION

These two schemes demonstrate methods for quantitative and qualitative risk assessment and evaluation for introduced species. Both of the examples illustrate how a common system of prediction and common valuation criteria can be applied across the full range of taxa that could be involved in an invasion. They produce as outputs either continuous risk distributions or stepped risk categories and they help policy makers to assess the range of uncertainty involved and the value of obtaining additional information that might reduce the level of uncertainty. In both cases the frameworks are intended as first-level screening tools that allow general comparisons of impact, rather than highly detailed tools for deciding the response to particular priority species. While these generalized approaches have limits, they provide a useful system for ranking risks. The significant advantage of a generic model for predicting impacts is that priorities can be determined by comparing expected impacts in a common monetary unit over similar timeframes. While the feasibility of such models has been demonstrated by the DEFRA research, more effort would be needed to obtain practical parameter estimates for the large number of potential pests that should be evaluated if they were to be used for comprehensive priority setting on a species-by-species basis. It would be most practical to use them in a general form to set priorities, and to limit their use in specific cases to particular invasive species when these have been detected or otherwise under special scrutiny.

REFERENCES

Defra, 2005. *Standard methodology to assess the risks from non-native species considered possible problems to the environment.* Department of Environment, Food and Rural Affairs, London. [www.defra.gov.uk/wildlife-countryside/resprog/findings/non-native-risks/]

FAO, 2003. *Pest risk analysis for quarantine pests, including analysis of environmental risks and living modified organisms.* Secretariat of the International Plant Protection Convention FAO, Rome. ISPM no. 11. [http://www.fao.org/DOCREP/006/Y4837E/Y4837E00.HTM]

Mumford, J.D., 2001. Environmental risk evaluation in quarantine decision making. *In:* Anderson, K., McRae, C. and Wilson, D. eds. *The economics of quarantine and the SPS agreement.* Centre for International Economic Studies, Adelaide, 353-383.

Mumford, J.D., 2002. Economic issues related to quarantine in international trade. *European Review of Agricultural Economics,* 29 (3), 329-348.

Mumford, J.D., 2005. Application of benefit/cost analysis to insect pest control in crops using the sterile insect technique. *In:* Dyck, V.A., Hendrichs, J. and Robinson, A.S. eds. *Sterile insect technique: principles and practice in area-wide integrated pest management.* Springer, Dordrecht, 481-498.

National Audit Office, 2003. *Protecting England and Wales from plant pests and diseases.* The Stationery Office, London. [http://www.nao.org.uk/publications/nao_reports/02-03/02031186.pdf]

OTA, 1993. *Harmful non-indigenous species in the United States.* Office of Technology Assessment, United States Congress, Washington.

Preston, C.D., Pearman, D.A. and Arnold, H.R. (eds.), 2002. *New atlas of the British and Irish flora: an atlas of the vascular plants of Britain, Ireland, the Isle of Man and the Channel Islands.* Oxford University Press, Oxford.

Standards Australia, 2004. *AS 4360 risk management portal.* Standards Australia, Sydney. [http://www.riskinbusiness.com]

Stohlgren, T.J. and Schnase, J.L., 2006. Risk analysis for biological hazards: what we need to know about invasive species. *Risk Analysis,* 26 (1), 163-173.

Waage, J.K., Fraser, R.W., Mumford, J.D., et al., 2004. *A new agenda for biosecurity.* Department for Environment, Food and Rural Affairs, London. [http://horizonscanning.defra.gov.uk/ViewDocument_Image.aspx?Doc_ID=173]

Waage, J.K., Mumford, J.D., Fraser, R.D., et al., 2005. Non-native pest species: changing patterns mean changing policy issues. *In: The BCPC congress: crop science & technology 2005.* British Crop Protection Council, Glasgow, 725-732.

Williamson, M., 1996. *Biological invasions.* Chapman & Hall, London. Population and Community Biology Series no. 15.

CHAPTER 13

ANALYSIS OF ENVIRONMENTAL RISKS

How to assess and manage risks of plants as pests?

GRITTA SCHRADER

Federal Biological Research Centre, Department for Plant Health, Messeweg 11/12, D-38104 Braunschweig, Germany. E-mail: g.schrader@bba.de

Abstract. Invasive alien plants can pose serious threats to cultivated and wild plants. This provides the basis to regulate them as 'plant pests' within the framework of plant health. To assess if a regulation would be appropriate, necessary and effective, and to identify available options for measures to reduce a possible risk, the revised International Standard on Phytosanitary Measures No. 11, "Pest risk analysis for quarantine pests including analysis of environmental risks and living modified organisms" by the International Plant Protection Convention or the more operational decision support scheme for pest risk analysis by the European and Mediterranean Plant Protection Organisation, present useful tools. One of the challenges to assess the risks of alien plants to other plants and the environment is the identification of the plant's potential for invasiveness. In addition, the approach to the economic impact assessment is different in comparison to the 'traditional' plant pests. The level of uncertainty is often greater in the assessment of environmental risks than in risks to cultivated plants, and also management options in particular for intentionally introduced plants can differ significantly from those for traditional pests. This article focuses on risk analysis beyond traditional plant quarantine, and elucidates the features with regard to the assessment and management of invasive alien plants.

Keywords: International Plant Protection Convention; plant health; invasive alien species; phytosanitary measures; quarantine pests

INTRODUCTION

The consideration of a plant as a pest is at first sight a quite unfamiliar point of view. And certainly, most plants would not fall into this category. But some plants that have been introduced into new ranges have shown invasive behaviour and are posing serious threats to cultivated and wild plants. In the majority of cases, these threats are caused by indirect damage that affects plants primarily by processes such as competition for space and resources or change of habitats, e.g., by altering soil chemistry or water regime.

According to the International Plant Protection Convention (IPPC) and as confirmed by FAO (1998), organisms that are directly or indirectly injurious to any kind of plants can be regulated within the framework of plant health, which aims to

A.G.J.M. Oude Lansink (ed.), New Approaches to the Economics of Plant Health, 191–202.
© 2007 *Springer*.

prevent introduction and spread of organisms harmful to plants and to promote appropriate measures for their control. Habitats and ecosystems can be protected from the consequences the introduction of an invasive alien plant may have, as they are essential for the survival of plants. Plant health is implemented in Europe by a long established and well developed system. Traditionally, only direct pests of plants (viruses, fungi, insects etc.) are regulated by this system, but the regulation of indirect pests – in particular invasive alien plants – is now under discussion.

The plant health system bases its phytosanitary measures on pest risk analysis (PRA), in order to assess whether an organism has a negative impact on plants and whether it should be regulated. The risk of introduction and spread of this pest is assessed and – if appropriate – options for measures are evaluated and proposed. Standards by the IPPC, in particular the International Standard on Phytosanitary Measures (ISPM) No. 11 "Pest risk analysis (PRA) for quarantine pests", and by the European and Mediterranean Plant Protection Organisation (EPPO) are available to facilitate this procedure. The EPPO standards, "Pest Risk Assessment Scheme" (Standard PM 5/3(1), EPPO 1997) and "Pest Risk Management Scheme" (Standard PM 5/4(1), EPPO 2000) were designed as user-friendly schemes to facilitate the conduct of PRA. Since they are based on the IPPC Standards, PRAs done with these schemes also provide – like PRAs based on the IPPC-PRA Standard itself – technical justification for the regulation of these organisms by states in the EPPO region. This is in accordance with the Sanitary and Phytosanitary Agreement under the World Trade Organisation (WTO 1994). Both IPPC and EPPO standards have recently been adapted to be better applicable to alien plants (for a background, see Schrader and Unger (2003)), because there are some significant differences in comparison to the 'traditional' plant pests. The revised ISPM No. 11 "Pest Risk Analysis for Quarantine Pests, Including Analysis of Environmental Risks and Living Modified Organisms" dating from 2003 is publicly available at the IPPC homepage (www.ippc.int). The two EPPO Standards for Pest Risk Assessment and Pest Risk Management have been combined to a single Standard on Pest Risk Analysis. This standard is fully in line with the revised IPPC Standard and will be available at the EPPO homepage (www.eppo.org) by the end of 2005. For the import of plants that are invasive or potentially invasive, another EPPO standard has been drafted addressing specific measures for such situations. This draft standard is currently under country consultation by the EPPO member states.

The objective of this paper is to describe an approach to evaluate the probability of the introduction and spread of invasive alien plants and the magnitude of the associated potential economic, including environmental, consequences. Differences between PRA for traditional pests and plants as pests are highlighted, and it is shown that management measures especially for intentionally introduced plants can differ significantly from those for direct pests of plants.

PLANTS AS PESTS

From a traditional plant pest, like a pathogen, a phytophagous insect or nematode, it is usually known beforehand that it can be harmful to plants, at least under certain conditions. For alien plants, the potential to cause damage is more the exception than the rule, and is often much more difficult to evaluate and to quantify. But in plant quarantine, an organism can only be regulated if it fulfils the criteria of a pest of plants and is of potential economic importance to the area endangered thereby (IPPC 1999). This includes by definition the consideration of environmental importance. A specific standard supplement (Glossary of Phytosanitary Terms, ISPM No. 5, Supplement No. 2, Guidelines on the Understanding of Potential Economic Importance and Related Terms Including Reference to Environmental Considerations from 2003) gives some details on this inclusion of environmental importance.

According to the Guiding Principles for the Prevention, Introduction and Mitigation of Impacts of Alien Species (CBD 2002) of the Convention on Biological Diversity (UNEP 1992), invasive alien species are non-indigenous organisms that threaten biodiversity. Consequently, an organism that solely poses a risk to crops or otherwise cultivated plants does not fall into the scope of the CBD. But an organism that does not have any adverse effect on crops or cultivated plants, can be considered a quarantine pest (definition: IPPC 1999), as long as there is a direct or indirect effect on other plants (ICPM 2001). An organism threatening biodiversity via an impact on plants fulfils the definitions for both an invasive alien species and a quarantine pest. Accordingly, all relevant threats to biodiversity as a consequence of the introduction and spread of organisms directly or indirectly harmful to plants are covered by the IPPC.

RISK ASSESSMENT OF ALIEN PLANTS

The first PRA standard by the IPPC, ISPM No. 2 "Guidelines for Pest Risk Analysis" from 1996 is currently under revision, describing the basic concept of pest risk analysis within the framework of the IPPC. It introduces the three stages of pest risk analysis – initiation, risk assessment and risk management – and the components for collecting, recording and communication of information. The initiation stage is explained in detail in this standard. Reference to other ISPMs, e.g. ISPM No. 11, is made regarding the risk assessment and risk management stages. The initiation stage aims at the recognition of organisms and pathways of phytosanitary concern that should be considered for pest risk assessment in relation to the identified PRA area. By risk assessment, the probability of the introduction and spread of a pest and the magnitude of the associated potential economic consequences are evaluated. With regard to alien plants, main questions are: does the plant have a high potential for spread and does it damage or threaten biodiversity? How is damage to be defined? Which effects does it have on other plants, on habitats, on ecosystems? In this context, it is important to define or estimate thresholds for invasiveness and environmental risks. Comparisons with other species, similar situations or experience from previous PRAs may help to answer these questions. Answers should be as objective and comprehensible as possible.

Introductions of alien plants into a country are intentional or unintentional and can be subjected to different motivations. For intentional introductions, the motivation is in most cases the trade with the plant itself, usually there is an interest on both sides, the exporter and the importer, to introduce this plant into a country. For unintentional introductions, the motivation is the trade with another subject, but this subject may be contaminated by plant propagules. Accordingly, some details for risk assessment differ for these two different situations.

Intentional introductions

In traditional PRA, it has to be assessed whether and how a pest can enter a country. For alien plants for planting, this is not necessary, as the entry is intended – such plants are intentionally imported, traded and planted. Instead, it is important to look at the pathway from the intended to the unintended habitat and the probability of establishment in the unintended habitat – in other words: can the plant escape from where it has been sown or planted? This assessment involves for example the consideration of climatic and other abiotic factors, the reproductive strategy of the plant species, the possibility of prevention of establishment by natural enemies or by competition from species present in the PRA area, and the likelihood of mitigation of impacts (eradication, containment or control) of the species after introduction.

Even the escape from an intended habitat is not causing any harm in probably most cases. A lot of plants are not able to establish permanently, others just blend into the environment without causing any problems. The tens rule (Williamson 1996), stating that 10 % of introduced species spread, 10 % of these establish, and 10 % of the established species cause problems (= 0.1 %), is tolerably applicable to alien plants. The difficulty is to identify this (approximately) 0.1 % that could be harmful among the plant species being introduced into a country or a region.

Examples for plants that have been introduced intentionally and that are causing problems now in Germany are Japanese knotweed (*Fallopia japonica*), American skunk-cabbage (*Lysichiton americanus*), giant hogweed (*Heracleum mantegazzianum*), golden rod (*Solidago canadensis* and *S. gigantea*) and black locust (*Robinia pseudoacacia*).

Unintentional introductions

Invasive alien plants may also be introduced unintentionally into a country as, e.g., contaminants of seeds, bird seed, oil seed, grain, fodder, wool, with soil or other growing medium, attached to vehicles or machines or in containers used for shipping. An example is annual ragweed (*Ambrosia artemisiifolia*), which is native to North America. This plant is introduced, e. g., with contaminated sunflower bird seed. Chufa flatsedge (*Cyperus esculentus*) has been unintentionally introduced and spread by vehicles and contaminated seed.

To reduce the risks of unintentional introductions of invasive alien plants, it is important to consider relevant pathways and to estimate the probability of the pest plant being associated with the pathway at origin. For example, in the case of *A. artemisiifolia*, it would be important to know if a sunflower field from which bird seed is produced is contaminated with plants of *A. artemisiifolia*, and if these plants are able to produce viable seeds. Also, it has to be considered if measures are applied to reduce or avoid contamination.

Also for unintentional introductions, the likelihood of spread and the potential to cause damage have to be assessed.

Assessment of the plant's potential for invasiveness

One of the challenges to assess the risks of alien plants is the identification of the plant's potential for invasiveness – this will probably in most cases be the major difficulty in the whole PRA. A key issue within PRA is to find out if the assessed organism has intrinsic attributes indicating that it could cause significant harm to plants or plant communities. Attributes of plants which could be relevant for invasiveness are broad ecological amplitude and high adaptability, ability to produce many seeds or vegetative propagules and to build up a persistent seed bank, and high competitive strength; see, for example, Rejmánek and Richardson (1996), Rejmánek (2000) and Heger and Trepl (2003). Important questions are also if the species is invasive in its area of current distribution, if the chances for rapid natural spread are high, if the propagules are highly mobile or if the plant benefits from cultivation or browsing pressure, and if there is a likelihood of building up monospecific stands. These attributes may increase the likelihood of invasion, but they are not in any case necessary for invasion success. On the other hand, a plant may have all these attributes without causing any problems. A big obstacle to general predictions is that there are no known broad scientific criteria for all (potentially) invasive plants in all relevant circumstances. Several publications deal with difficulties related to prediction of invasiveness (e.g. Williamson and Brown 1986; Kolar and Lodge 2001; Williamson 2001; Heger and Trepl 2003). According to Williamson (1999) invasiveness elsewhere is a comparably consistent predictor.

To get a better prediction of the plant's ability to invade, experimental plantings could be an option, but the time-lag effect is difficult to be assessed. An invasion is often triggered by planting large volumes of a plant species, and by repeated and secondary introductions (see e.g. Kowarik 2003) – therefore, intended use and volume of the introduction of a plant species need to be taken into account as well. The success of a plant in invading a certain area will also depend on the susceptibility to invasion of the related habitat, so this has to be considered too (Heger and Trepl 2003). Often, sites modified by man or strongly disturbed grounds are especially prone to invasion. With regard to the alien plant's requirements, nutrient-rich or nutrient-poor soils (or waters) may be preferred, and some plants, like *Lysichiton americanus*, are predominantly found in vegetation close to nature (Alberternst and Nawrath 2002).

Despite all these difficulties, a screening for first-time introductions of plant species would be useful, with simple criteria, especially invasiveness elsewhere, followed by an in-depth risk analysis in case there is some indication for invasiveness.

Primary and secondary consequences resulting from establishment and spread

If a species has been identified to be invasive or potentially invasive, the next step is to specify the concrete possible consequences of establishment and spread. In this context, primary and secondary consequences have to be considered. Regarding environmental risks, important primary consequences would be for example the reduction of the abundance of keystone plant species, which are 'responsible' for the existence of an ecosystem of a certain type, or are the main drivers for the development of or succession within an ecosystem. Also, for species that are major components of ecosystems, the decision may be taken that they should be protected, because reduction of their abundance will certainly change the habitat or ecosystem that is dependent on them and this change is not desired. This would especially be the case if their reduction causes the ecosystem to degrade or to collapse. Negative impacts on endangered native plant species must also be prevented in order to protect biodiversity. Furthermore, protection of other plant species against significant reduction, displacement or elimination is taken into account, though endangered species receive more attention than just 'normal' species because of their status.

Examples for secondary consequences relate to significant effects on plant communities, significant effects on designated environmentally sensitive or protected areas, significant changes in ecological processes and of the structure, stability or dynamics of an ecosystem (including further effects on plant species, erosion, water-table changes, increased fire hazard, nutrient cycling, etc.), effects on human use (e.g., water quality, recreational uses, tourism, animal grazing, hunting, fishing), or costs of environmental restoration. If for example the symbiotic nitrogen-fixing black locust (*Robinia pseudoacacia*) is invading certain habitats it may have a significant effect on the whole plant community, because ecological processes may be affected by an accumulation of nutrients due to nitrogen enrichment of the soil. This has a significant negative impact on nutrient-poor soils, which often are habitats for endangered plant species. In invaded regions in Poland, enrichment of soil nitrogen by *R. pseudoacacia* is thought to favour the appearance of certain combinations of associated species (Dzwonko and Loster 1997).

Yet another example is the damage that could be caused by the aquatic plant New Zealand pygmyweed (*Crassula helmsii*). Its vegetative growth leads to dense mats that can block ponds and drainage ditches, and even outcompete native flora and impoverish the ecosystem for invertebrates and fish. The vegetation mats can be dangerous to pets, livestock and children who confound them with dry land. *Hydrocotyle ranunculoides*, another aquatic plant, may change aquatic habitats by excluding light from the water, reducing photosynthesis to a significant extent.

Other negative impacts of introduced invasive alien plants can be allelopathic effects or hybridization. *Ailanthus altissima*, for example, has allelopathic effects on many other tree species and may consequently inhibit succession (Heisey 1996). Alien species can hybridize with closely related natives, which may lead to a loss of genetic and species diversity. An example is the American grass species *Spartina alternifolia*, which was accidentally introduced and hybridized with *S. maritima* in Britain, producing *S. x townsendii*. The hybrid led to a tetraploid species, *S. anglica*, which outcompeted the parent species and is invading successfully British wetlands (Gray et al. 1991; Thompson 1991). Alien plant species may also hybridize with other non-natives, possibly leading to the evolution of a stronger, more vigorous hybrid. This has been observed with the hybridization of *R. japonica* and *R. sachalinensis*. resulting in the highly invasive hybrid *Reynoutria x bohemica* (Pyšek et al. 2003).

Assessment of economic consequences

Estimations of consequences of introduction, establishment and spread of an alien plant made up to this point, related to the hypothetical situation that it has been introduced and fully expresses its economic consequences in the PRA area. But in practice, economic consequences are related to time and place factors. The total economic consequences for more than a year can be expressed as net present value of annual economic consequences, and an appropriate discount rate selected to calculate net present value. Economic consequences also depend on speed of spread, on the number of habitats infested and a change of relevant factors over time.

Environmental effects can be of an economic nature, without having an existing market that can be easily identified. Therefore, the effects may not be adequately measured in terms of prices in established product or service markets. These impacts could be approximated with an appropriate non-market valuation method. The assessment of consequences may be quantitative or qualitative. Often, qualitative data are sufficient. A quantitative method may not exist to address a situation (e.g., catastrophic effects on a keystone species), or a quantitative analysis may not be possible (no methods available).

For a valuation of the environment, ISPM No. 11 provides different methodologies, including the consideration of 'use' and 'non-use' values. 'Use' values can be separated into consumptive (e.g., fishing in a lake) and non-consumptive (e.g., using forests for leisure activities). 'Non-use' values can be divided into option value (value for use at a later date), existence value (knowledge that an element of the environment exists), and bequest value (knowledge that an element of the environment is available for future generations). To assess these values, methods are available referring to market-based approaches, surrogate markets, simulated markets, and benefit transfer. Also, the assessment can be based on non-monetary valuations, such as number of species affected or water quality. In any case, these methodologies are best applied in consultation with experts in economics. The procedures should be documented, consistent and transparent and environmental values should be clearly categorized.

Uncertainty

The level of uncertainty is often greater in the assessment of risks to the environment than of risks to cultivated plants, due to the lack of information, additional complexity associated with ecosystems, and variability associated with pests, hosts or habitats. Generally, phytosanitary measures are intended to account for uncertainty but should not be more stringent than necessary. For the identification of management options it is important to consider the degree of uncertainty.

RISK MANAGEMENT OF ALIEN PLANTS

If the assessment of an organism for which the PRA is being done reveals an unacceptable risk to plants in the PRA area, management options have to be identified to reduce or exclude these risks. The situation with unintentionally introduced invasive alien plants (as, for example, propagules or hitchhikers with other plants) is comparable with other plant pests – measures may be determined that block or reduce entry and spread via the identified pathway(s). But with intentionally introduced plants, management options are quite different. The management part of ISPM No. 11 does not give detailed guidance on how to proceed with the import of invasive or potentially invasive plants. In the framework of EPPO, a standard for the import of alien plants has therefore been drafted that will most certainly be adopted by the EPPO Council in autumn 2006. Important points to consider are: the surveillance after planting, the preparation of control or emergency plans if a plant is found outside its intended habitat and spreads to an unacceptable degree, the restriction on import, sale, holding and on planting (including authorization of intended habitats, prohibition of planting in unintended habitats, required growing conditions for plants), the notification before import, restrictions on movement (e.g., prevention of movement to specified areas), and the obligation to report findings. In any case, the intended use of the plant is influencing the choice of management measures. A differentiation between the intended use of species, e.g., for gardening (within urban areas) or for landscaping (planted in large numbers, at many different locations, in the countryside) can also influence the selection of possible measures. The decision if such measures have to be applied to certain imported plants needs to be based on a pest risk assessment. A quick screening of the plant should indicate whether a detailed risk assessment is necessary. If this is the case, and an unacceptable risk is identified, the most appropriate measures as indicated in the draft EPPO standard could be selected.

For plants new to a territory, it is difficult to predict their ability to invade. If an invasive behaviour has never been observed before but some characteristics or attributes of the plants and their potential habitats raise suspicion for invasiveness, an option could be not to take phytosanitary measures at import, but to apply surveillance or other procedures after entry and to monitor plants after import and planting. The decision to select the most adequate approach has to be based on expert judgement – usually, this would be the risk assessor. This could be combined with an emergency plan to be used when the plant is found outside its intended habitats in undesirable numbers. Also, the phenomenon of 'time lag' has to be considered – some invasive species, especially plants, only show invasive behaviour after a considerable time.

As measures for ornamental plants may be difficult to understand for the public, raising of publicity is an important point in this context. Measures may easier be accepted for clear-cut cases than for plants for which only a risk potential has been identified.

REGULATORY FRAMEWORK FOR INVASIVE ALIEN SPECIES AFFECTING PLANTS

The international regulatory framework for organisms that are directly or indirectly injurious to any kind of plants is provided by the International Plant Protection Convention (IPPC), which is an international treaty adopted in 1952 and revised in 1979 and 1997 (IPPC 1999). Its aim is to secure action to prevent the spread and introduction of pests of plants and plant products, and to promote appropriate measures for their control. Promoted by an increased awareness for the protection of the environment, the IPPC has started in 1999 to identify explicitly how its implementation directly relates to the identification of environmental risks caused by plant pests. Although the IPPC addresses the spread of pests associated with international trade, the Convention is not limited in this respect – it is focussed on the protection of plants in general. This includes the protection of biodiversity, and many provisions, procedures and standards of the IPPC are directly relevant to, or overlap with, the aim of article 8 (h) of the Convention on Biological Diversity (CBD), which requires contracting parties to "prevent the introduction of, control or eradicate those alien species which threaten ecosystems, habitats or species" (UNEP 1992). In order to clarify the role and competence of the IPPC, to avoid overlaps and double work, and to achieve a synergistic approach regarding the protection against invasive alien species with impacts on plants, the IPPC works collaboratively with the CBD. A memorandum of cooperation between IPPC and CBD has been adopted in 2004.

Regional Plant Protection Organizations provide coordination on a regional level for the activities and objectives of the IPPC and help contracting parties meet the Convention's obligations. The Regional Plant Protection Organization for Europe, Russia and several other countries from the former Sowjet Union, as well as some countries of the Middle East and North Africa, is the European and Mediterranean Plant Protection Organisation (EPPO). It was founded in 1951 with its own convention. Following to the activities of the CBD and the IPPC (see, e.g., Schrader and Unger 2003), the European and Mediterranean Plant Protection Organisation is developing a new working program on invasive alien species and 'pest plants'. EPPO gives recommendations to its 47 member countries on how to assess and manage invasive alien plants. Plants are listed on the new EPPO list for invasive alien plants (consisting of 34 species at the moment; http://www.eppo.org/QUARANTINE/ias_plants.htm) or on the EPPO action list (2 species at the moment: *Hydrocotyle ranunculoides* and *Lysichiton americanus*; http://www.eppo.org/QUARANTINE/action_list.htm).

Phytosanitary measures in the European Union have been fully harmonized in 1993, when the EU internal market was established. With EU Council Directive 2000/29/EC (European Commission 2000) protective measures against the introduction of organisms harmful to plants or plant products into the EU Member

States from other EU Member States or from third countries are regulated. Similarly, protective measures against the spread of harmful organisms within the Community through movements of plants, plant products and other related objects within a Member State are included.

One of the most important measures in this Directive is the listing of harmful organisms whose introduction into the community must be prohibited. The related annexes to the Directive list quarantine pests as well as plants, plant products and other articles that could be pathways for these quarantine pests. Their introduction and movement into or within the EU are prohibited or subject to certain requirements or restrictions. These annexes contain binding measures for more than 200 organisms. Currently, mainly pests directly harmful to cultivated plants, like insects, nematodes and viruses, are listed. Many of these organisms also pose a threat to biodiversity. Invasive alien plants have not been enclosed in these annexes up to now – besides some non-European parasitic plants of the genus of *Arceuthobium* – but their inclusion is presently under discussion. First candidates are *Hydrocotyle ranunculoides* and *Lysichiton americanus*, which are already listed at the EPPO Action list.

In addition to these 'black lists', implementing provisions may be adopted to lay down conditions for the introduction into the Member States and the spread within the Member States of organisms that are suspected of being harmful to plants or plant products but are not listed.

CONCLUSIONS

In this paper it has been shown that different aspects are relevant for the assessment and management of invasive alien plants in comparison to traditional pests of plants. As plants are usually not pests or do not behave like pests in their area of origin, it is often difficult to predict their potential risks to other plants if introduced to new areas. Management measures have to be selected in proportion to the risk – the higher the risk, the more stringent a measure.

For invasive alien plants that threaten other plants or plant products and for the analysis of environmental risks, the revised IPPC and EPPO standards on PRA provide the necessary elements for a substantial risk analysis. As these tools have originally been used for traditional pests and have only recently been revised to be better applicable to the assessment and management of environmental risks, experience for their application and the implementation of their results in this regard has yet to be increased. However, contracting parties (number as of 14 October 2005: 141) to the IPPC benefit from having in place long-established IPPC-based systems for the prevention of introduction and spread of plant pests and can abstract experience from this source to environmental risks. Especially the assessment of risks posed by alien plants to a PRA area is a difficult task because of high levels of complexity in ecosystems, uncertainty about threats to biodiversity, pressure arising from globalization including trade and tourism, etc.

Results of the PRAs can be used for recommendations by EPPO to its Member Countries, including proposals for management options. PRAs and EPPO management options could provide the basis for the EU Commission and accordingly for separate EPPO Member Countries that are not EU Member States to regulate specified invasive alien plants, including prohibition of import or conditions for introduction or use.

REFERENCES

Alberternst, B. and Nawrath, S., 2002. *Lysichtion americanus* Hulten & St. John neu in Kontinental-Europa. Bestehen Chancen für die Bekämpfung in der Frühphase der Einbürgerung? *In:* Kowarik, I. and Starfinger, U. eds. *Biologische Invasionen: eine Herausforderung zum Handeln?* 91-99. Neobiota no. 1.

CBD, 2002. Decision VI/23. Alien species that threaten ecosystems, habitats or species. II Guiding principles for the prevention, introduction and mitigation of impacts of alien species. *In: COP 6 Decisions: decisions adopted by the Conference of the Parties to the Convention on Biological Diversity at its sixth meeting, The Hague, 7-19 April 2002.* Convention on Biological Diversity, Montreal, 249-261. [http://www.biodiv.org/decisions/default.aspx?m=COP-06&id=7197&lg=0]

Dzwonko, Z. and Loster, S., 1997. Effects of dominant trees and anthropogenic disturbances on species richness and floristic composition of secondary communities in southern Poland. *Journal of Applied Ecology,* 34 (4), 861-870.

EPPO, 1997. *Pest risk assessment scheme (Standard PM 5/3 (1)).* European and Mediterranean Plant Protection Organization.

EPPO, 2000. *Pest risk management scheme (Standard PM 5/4(1)).* European and Mediterranean Plant Protection Organization.

European Commission, 2000. EC Council Directive 2000/29/EC of 8 May 2000 on Protective measures against the introduction into the Community of organisms harmful to plants or plant products and against their spread within the Community. *Official Journal of the European Communities,* L 169/1 (10.7.2000), 1-112 and subsequent amendments. [http://europa.eu.int/eur-lex/pri/en/oj/dat/2000/l_169/l_16920000710en00010112.pdf]

FAO, 1998. Appendix I. Interpretations as agreed by the fourteenth session of the Committee on Agricuture (7-11 April 1997). *In: Excerpts from the Report of the Conference of FAO (C97/REP), twenty-ninth session, Rome, 7-18 November 1997.* FAO, Interim Commission on Phytosanitary Measures, Rome. Publication no. ICPM-98/INF/1.

FAO, 2003. *Pest risk analysis for quarantine pests, including analysis of environmental risks and living modified organisms.* Secretariat of the International Plant Protection Convention FAO, Rome. ISPM no. 11. [http://www.fao.org/DOCREP/006/Y4837E/Y4837E00.HTM]

FAO, 2006. *International standards for phytosanitary measures: glossary of phytosanitary terms.* Secretariat of the International Plant Protection Convention FAO, Rome. ISPM no. 05. [https://www.ippc.int/servlet/BinaryDownloaderServlet/133607_ISPM05_2006_E.pdf?filename=115 1504714760_ISPM05_2006_E.pdf&refID=133607]

Gray, A.J., Marshall, D.F. and Raybould, A.F., 1991. A century of evolution in *Spartina anglica. Advances in Ecological Research,* 21, 1-62.

Heger, T. and Trepl, L., 2003. Predicting biological invasions. *Biological Invasions,* 5 (4), 313-321.

Heisey, R.M., 1996. Identification of an allelopathic compound from *Ailanthus altissima* (Simaroubaceae) and characterization of its herbicidal activity. *American Journal of Botany,* 83 (2), 192-200.

ICPM, 2001. Appendix XIII. Statements of the ICPM exploratory open-ended working group on phytosanitary aspects of genetically modified organisms, biosafety and invasive species. *In: Report of the Third Interim Commission on Phytosanitary Measures, 2-6 April, 2001, Rome, Italy.* FAO, Interim Commission on Phytosanitary Measures, Rome. [https://www.ippc.int/servlet/BinaryDownloaderServlet/14320_ICPM_Report_2001___E.PDF?filename=1079019159579_ICPM3 e.PDF&refID=14320]

IPPC, *IPPC Standards available at the International Phytosanitary Portal.* International Plant Protection Convention. [https://www.ippc.int/].

IPPC, 1999. *International Plant Protection Convention: New revised text approved by the FAO Conference at its 29th session, November 1997.* FAO, Rome. [https://www.ippc.int/servlet/ BinaryDownloaderServlet/13742_1997_English.pdf?filename=/publications/13742.New_Revised_T ext_of_the_International_Plant_Protectio.pdf&refID=13742]

Kolar, C.S. and Lodge, D.M., 2001. Progress in invasion biology: predicting invaders. *Trends in Ecology and Evolution,* 16 (4), 199-204.

Kowarik, I., 2003. *Biologische Invasionen: Neophyten und Neozoen in Mitteleuropa.* Ulmer, Stuttgart.

Pyšek, P., Brock, J.H., Bímová, K., et al., 2003. Vegetative regeneration in invasive *Reynoutria* (Polygonaceae) taxa: the determinant of invasibility at the genotype level. *American Journal of Botany,* 90 (10), 1487-1495.

Rejmánek, M., 2000. Invasive plants: approaches and predictions. *Australian Ecology,* 25 (5), 497-506.

Rejmánek, M. and Richardson, D.M., 1996. What attributes make some plant species more invasive? *Ecology,* 77 (6), 1655-1661.

Schrader, G. and Unger, J.G., 2003. Plant quarantine as a measure against invasive alien species: the framework of the International Plant Protection Convention and the plant health regulations in the European Union. *Biological Invasions,* 5 (4), 357-364.

Thompson, J.D., 1991. The biology of an invasive plant. What make *Spartina anglica* so successful? *Bioscience,* 41 (6), 393-401.

UNEP, 1992. *Convention on biological diversity: text and annexes.* United Nations Environment Programme, Rio de Janeiro.

Williamson, M., 1996. *Biological invasions.* Chapman & Hall, London. Population and Community Biology Series no. 15.

Williamson, M., 1999. Invasions. *Ecography,* 22 (1), 5-12.

Williamson, M., 2001. Can the impacts of invasive plants be predicted? *In:* Brundu, G., Brock, J., Camarda, I., et al. eds. *Plant invasions: species ecology and ecosystem management.* Backhuys Publishers, Leiden, 11-19.

Williamson, M.H. and Brown, K.C., 1986. The analysis and modelling of British invasions. *Philosophical Transactions of the Royal Society of London. Series B. Biological Sciences,* 314 (1167), 505-522.

WTO, 1994. *SPS-Agreement:agreement on the application of sanitary and phytosanitary measures.* World Trade Organisation, Geneva.

Wageningen UR Frontis Series

1. A.G.J. Velthuis, L.J. Unnevehr, H. Hogeveen and R.B.M. Huirne (eds.): *New Approaches to Food-Safety Economics.* 2003
   ISBN 1-4020-1425-2; Pb: 1-4020-1426-0
2. W. Takken and T.W. Scott (eds.): *Ecological Aspects for Application of Genetically Modified Mosquitoes.* 2003
   ISBN 1-4020-1584-4; Pb: 1-4020-1585-2
3. M.A.J.S. van Boekel, A. Stein and A.H.C. van Bruggen (eds.): *Proceedings of the Frontis workshop on Bayesian Statistics and quality modelling.* 2003
   ISBN 1-4020-1916-5
4. R.H.G. Jongman (ed.): *The New Dimensions of the European Landscape.* 2004
   ISBN 1-4020-2909-8; Pb: 1-4020-2910-1
5. M.J.J.A.A. Korthals and R.J.Bogers (eds.): *Ethics for Life Scientists.* 2004
   ISBN 1-4020-3178-5; Pb: 1-4020-3179-3
6. R.A. Feddes, G.H.de Rooij and J.C. van Dam (eds.): *Unsaturated-zone modeling.* Progress, challenges and applications. 2004   ISBN 1-4020-2919-5
7. J.H.H. Wesseler (ed.): *Environmental Costs and Benefits of Transgenic Crops.* 2005   ISBN 1-4020-3247-1; Pb: 1-4020-3248-X
8. R.S. Schrijver and G. Koch (eds.): *Avian Influenza.* Prevention and Control. 2005
   ISBN 1-4020-3439-3; Pb: 1-4020-3440-7
9. W. Takken, P. Martens and R.J. Bogers (eds.): *Environmental Change and Malaria Risk.* Global and Local Implications. 2005
   ISBN 1-4020-3927-1; Pb: 1-4020-3928-X
10. L.J.E.J. Gilissen, H.J. Wichers, H.F.J. Savelkoul and R.J. Bogers, (eds.): *Allergy Matters.* New Approaches to Allergy Prevention and Management. 2005
    ISBN 1-4020-3895-X; Pb: 1-4020-3896-8
11. B.G.J. Knols and C. Louis (eds.): *Briding Laboratory and Field Research for Genetic Control of Disease Vectors.* 2006
    ISBN 1-4020-3800-3; Pb: 1-4020-3799-6
12. B. Tress, G. Tress, G. Fry and P. Opdam (eds.): *From Landscape Research to Landscape Planning.* Aspects of Integration, Education and Application. 2006
    ISBN 1-4020-3979-4; Pb: 1-4020-3978-6
13. J. Hassink and M. van Dijk (eds.): *Farming for Health.* Green-Care Farming Across Europe and the United States of America. 2006
    ISBN 1-4020-4540-9; Pb: 1-4020-4541-7
14. R. Ruben, M. Slingerland and H. Nijhoff (eds.): *The Agro-Food Chains and Networks for Development.* 2006   ISBN 1-4020-4592-1; Pb: 1-4020-4600-6
15. C.J.M. Ondersteijn, J.H.M. Wijnands, R.B.M. Huiren and O. van Kooten (eds.): *Quantifying the Agri-Food Supply Chain.*
    ISBN 1-4020-4692-8; Pb: 1-4020-4693-6
16. M. Dicke and W. Takken (eds.): *Chemical Ecology.* From Gene to Ecosystem. 2006   ISBN 1-4020-4783-5; Pb: 1-4020-4792-4
17. R.J. Bogers, L.E. Craker and D. Lange (eds.): *Medicinal and Aromatic Plants.* Agricultural, Commercial, Ecological, Legal, Pharmacological, and Social Aspects. 2006   ISBN 1-4020-5447-5; Pb: 1-4020-5448-3

18. A. Elgersma, J. Dijkstra and S. Tamminga (eds.): *Fresh Herbage for Dairy Cattle. The Key to a Sustainable Food Chain.* 2006

ISBN 1-4020-5450-5; Pb: 1-4020-5451-3

19. *To be published.*

20. A.G.J.M. Oude Lansink (ed.): *New Approaches to the Economics of Plant Health.* 2007 ISBN 1-4020-5825-X; Pb: 1-4020-5826-8